GERMAN WEAPONS OF WORLD WAR II

GERMAN WEAPONS OF WORLD WAR II

Stephen Hart

amber BOOKS

Published by
Amber Books Ltd
United House
North Road
London N7 9DP
United Kingdom
www.amberbooks.co.uk
Instagram: amberbooksltd
Facebook: www.facebook.com/amberbooks
Twitter: @amberbooks

ISBN: 978-1-78274-629-4

Project Editor: Michael Spilling
Designer: Brian Rust
Picture Researcher: Terry Forshaw

Printed in China

ILLUSTRATION CREDITS

PHOTOGRAPHS

Art-Tech: 39, 45, 46, 58, 68, 70, 73, 82, 91, 107, 108, 112, 124–195, 209, 211
Cody Images: 7, 8, 20, 24, 28, 34, 56, 75, 102, 106, 113–120, 198–206, 212–216
Nik Cornish/STAVKA: 71, 89, 96
Ukrainian State Archive: 15, 100
US National Archives: 36, 41

ARTWORKS

Alcaniz Fresno's: 10, 11, 13–22, 29, 30, 32, 33, 37, 38, 44–49, 51, 52, 57–65, 67, 85, 86, 88, 92, 95, 101
Art-Tech: 12, 25, 27, 31, 35, 40, 42, 50, 53, 55, 71–84, 87, 90, 94, 97, 99, 104–219
Oliver Missing (www.o5m6.de): 66

Contents

Introduction

Many factors explain the stunning German military successes of 1939–41, as well as the dogged defensive determination witnessed during 1943–45 as the Allied advance rolled back the frontline of German-occupied Europe. Undoubtedly doctrine, tactics and leadership played a part in both German offensive success and defensive resilience. Yet the effective weapons and equipment the German armed forces employed played a critical role during this titanic six-year clash of arms.

The stunning German victories in Poland, Norway, France, and the Soviet Union from 1939–41 owed much to the effective all-arms and inter-service co-operation achieved. Panzer II–IV tanks, Sdkfz 250 APCs carrying Ka 98 rifle-armed grenadiers, 10.5cm leFH 18 howitzers and Ju 87 Stuka dive-bombers co-operated to deliver overwhelming combat power. Aerial superiority, like that secured during the May-June 1940 Western campaign, enabled the Luftwaffe to assist the unfolding ground offensive. Stuka dive-bombers provided contact-battle fire support; Do 17s delivered interdiction strikes; and Ju 52s inserted paratroopers behind enemy lines and mounted resupply sorties.

As the strategic advantage in the war shifted during 1942–43, the German armed forces thereafter displayed dogged defensive determination as the Allies gradually fought their way towards the Reich. This defensive resilience owed much to effective new weaponry, including: the MG 42 machine-gun and StG 44 assault-rifle; the heavily-armed and well-protected Tiger tank; the potent 8.8cm Pak 43 anti-tank gun; and the Fw 190-G ground-attack aircraft.

By the war's end, the Germans had developed many potent cutting-edge weapons, but these could only be deployed in insufficient numbers or with inadequate support to exert any significant impact on the war. Meanwhile, the strategic war against

Russian Front
With water streaming off its tracks, a German Panzer III from the 18th Panzer Division powers up a steep river bank having forded the River Bug during the German invasion of the Soviet Union, June 1941.

Allied shipping – spearheaded by Kriegsmarine capital ships and submarines – represented a key dimension of the war. Ultimately by 1944, the Germans had lost the Battle of the Atlantic and their early 1945 initiation of the new potent 'Electro-boat' submarine offensive came too late to delay Germany's imminent strategic defeat. That these weapons could even be deployed owed everything to the German war economy's developmental and manufacturing activities. During 1942–45, this was subjected to increasingly devastating attack by Allied strategic bombers. The defence of the Reich correspondingly became a critical mission. Thousands of anti-aircraft guns, like the 8.8cm Flak 41, joined Fw 190, Me 110 and Me 410 fighter-interceptors in striving to shoot down Allied bombers before they dropped their payloads onto the intended target. During 1944–45, a new generation of impressive jet fighters – like the Me 262 and He 162 – joined this struggle, but in insufficient numbers to delay Germany's looming strategic defeat.

Thus, the course and outcome of World War II was significantly determined by the weapons and equipment the German armed forces developed, improved and employed during this titanic conflict.

Chapter 1

Tracked and Wheeled Vehicles

Germany's stunning 1939–41 military victories owed much to the effective weapons and equipment that its armed forces employed when wedded to appropriate doctrine, tactics and leadership.

The armoured fighting vehicle (AFV) that most epitomized the fast-paced strategic armoured warfare termed *Blitzkrieg* ('Lightning War') was the tank (or 'panzer'). Combining lethality, protection and mobility, the tank was central to German *Blitzkrieg* operations.

TANKS

The Germans started the war in 1939 with six types of light/medium tanks: the Panzer I, II, 35(t), III, 38(t) and IV. Like all other nations, German tank design developed swiftly in the crucible of total war.

Of the extant six models, the Germans improved the Panzer III and IV, lengthening their service life, although by late 1943 the former was outgunned by its Allied adversaries. Simultaneously, the Germans developed new and ever more potent tanks, such as the Panzer V Panther medium vehicle, and the Panzer VI Tiger and VIB King Tiger heavy tanks.

Opposite:
Ardennes Offensive
Panther tanks move along a snow-covered road somewhere in the Ardennes region of Belgium during the Ardennes Offensive, December 1944. These tanks probably belong to the Sixth Panzer Army.

Panzer I Ausf B

The Panzer I Model B differed from its predecessor by having a slightly elongated suspension that incorporated a fifth medium-sized, open-spoked road wheel and a more raised rear idler wheel.

Panzerkampfwagen I Ausf B

Crew: 2
Production: 1934–1938
Weight: 8.9 tonnes (8.8 tons)
Dimensions: length: 4.8m (15ft 8in); width: 2.22m (7ft 4in); height: 1.99m (6ft 6in)
Engine: 44kW (60hp) Krupp M 305 four-cylinder air-cooled gasoline engine
Speed: 40km/h (25mph)
Range: 200km (124 miles)
Armament: 2 × 7.92mm (0.31in) MG13 machine guns
Armour: 7–13mm (0.28–0.5in)

Panzer I light tank

In 1932, the German Army required a stopgap light tank to train armoured troops, while more potent vehicles (the Panzer III and IV) were developed. This vehicle was designated Agricultural Tractor Model I (LaS I) to conceal the tank production programme, which contravened the Treaty of Versailles; it was subsequently redesignated Panzer I (Panzerkampfwagen I or Armoured Fighting Vehicle I).

In 1934–35, Krupp manufactured 300 LaS IA tanks; these two-man, 5.4-tonne (5.3-ton) vehicles possessed 6–13mm (0.24–0.5in) thick armour and were powered by a 42kW (57hp) engine. During 1935–39, Krupp produced 1203 heavier up-engined LaS IB tanks; these had an elongated suspension and a fifth road wheel.

The Panzer I first saw combat with the German Condor Legion in the Spanish Civil War. During the September 1939 German invasion of Poland, 1445 Panzer Is fought alongside 1532 Panzer II, III and IV tanks. Despite the stunning victory, the Germans lost 89 lightly armoured Panzer Is, mostly to enemy fire. Desultory Panzer III and IV production meant that the Germans still had to commit 619 Panzer I tanks to their May 1940 invasion of the West, although merely in a scouting role. Despite this, losses were high and Germany never again employed the Panzer I as a combat tank. Instead, the Germans either employed the 800 remaining Panzer Is in training/garrison duties in Nazi-occupied Europe or converted them into armoured support vehicles.

Panzer II light tank

In the early 1930s, the Germans developed the Panzer II, a heavier version of the Panzer I, while Panzer III and IV development unfolded. Mass production of the Panzer II Model A commenced in 1937. Powered by a 104kW (140hp), engine this lightly armoured 8.9-tonne (8.8-ton) tank mounted twin 20mm (0.79in) cannon.

In 1937–39, the slightly modified Models B and C entered service, the latter incorporating an up-armoured driver's vision port. Subsequently, during 1938–39, 250 Models D and E fast reconnaissance tanks entered service.

These fast tanks possessed modified tracks and a novel torsion-bar suspension that boosted maximum road speed to 56km/h (35mph). Some 1223 Panzer II Models A–E tanks fought in the autumn 1939 German invasion of Poland.

In March–December 1941, 233 Model F tanks were produced. The heavy losses suffered during the April–June 1940 campaigns against Norway and France indicated that the Panzer II required greater survivability. The Model F thus possessed 35mm (1.38in) frontal and 30mm (1.18in) side armour; the increased weight reduced this variant's top speed to 40km/h (25mph). The experience of the 1941 Soviet campaign revealed that the Panzer II was nearing the end of its active combat life.

However, the Germans continued to produce Model G and J variants, featuring external turret stowage bins, well into 1942. In late 1942, the Germans gradually phased the Panzer

Panzerkampfwagen II Ausf F
Crew: 3
Production: 1934–1936
Weight: 8.9 tonnes (8.8 tons)
Dimensions: length: 4.81m (15ft 10in); width: 2.22m (7ft 4in); height: 1.99m (6ft 6in)
Engine: 103kW (138hp) Maybach HL62TR
Speed: 40km/h (25mph)
Range: 200km (124 miles)
Armament: 1 × 2cm (0.79in) KwK 30
Armour: 5–35mm (0.19–1.38in)

Panzer II, Afrika Korps
The white palm tree insignia on this vehicle's front hull side and its desert camouflage scheme identify this late-production Panzer II as one deployed by the Afrika Korps in North Africa.

II out of combat and used its chassis for other armoured support vehicles.

Panzer 35(t) light tank

In 1939, the Germans took over the Czech Army's equipment, including 218 LT35 medium tanks, which were redesignated Panzer 35(t)s. This four-man, 10.5-tonne (10.3-ton) tank was protected by 35mm (1.38in) thick frontal armour and mounted a 37.2mm (1.46in) Skoda gun. With a suspension based on eight paired leaf-spring road wheels, the tank had a top road speed of 40km/h (25mph). All 218 Panzer 35(t)s participated in the September 1939 Axis invasion of Poland, with 112 serving in the 1st Light Division. During the division's 800-km (497-mile) advance on Radom, the Panzer 35(t) proved its lethality, mobility and mechanical reliability, while displaying that it needed extensive maintenance. Skoda manufactured 31 more Panzer 35(t)s in 1939–40. Some 204 Panzer 35(t) tanks participated in the May 1940 German invasion of the West.

Some 162 Panzer 35(t)s took part in the 1941 German invasion of the Soviet Union, advancing 810 miles (1304km) before suffering particularly heavy losses around Klin, near Moscow. As winter set in, the Germans discovered that the tank's pneumatic steering system froze easily. By spring 1942, moreover, the Panzer 35(t) had become obsolete, so the Germans withdrew from front line service the few tanks still deployed on the Eastern Front. One last batch of Panzer 35(t)s fought during the Soviet winter 1942–43 counter-offensive with the newly committed 22nd Panzer

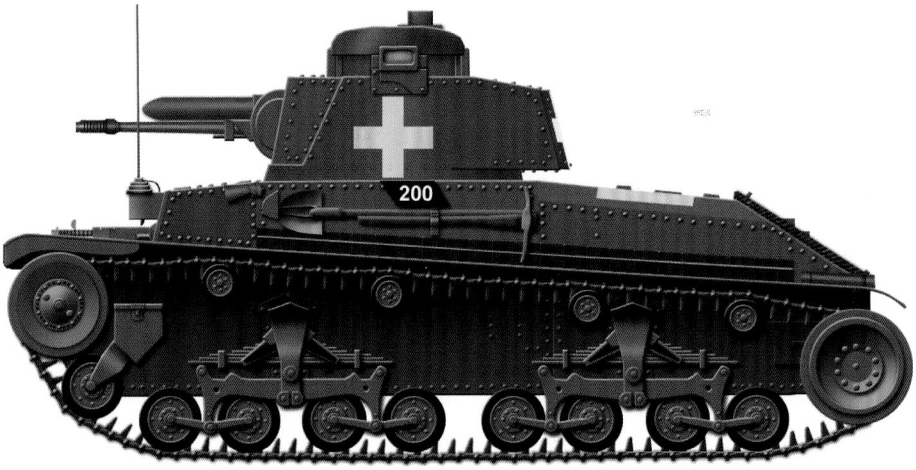

Panzer 35(t)
The distinctive design of the 3.72cm Skoda A3 gun, the tall domed commander's cupola on the turret roof, and the riveted armour help identify this vehicle as a Panzer 35(t) tank.

Panzerkampfwagen 35(t)
Crew: 4
Production: 1936–1940
Weight: 10.5 tonnes (10.3 tons)
Dimensions: length: 4.9m (16ft 1in); width: 2.06m (6ft 10in); height: 2.37m (7ft 10in)
Engine: 89kW (120hp) 4-cylinder, water-cooled Škoda T11/0 gasoline
Speed: 34km/h (21mph)
Range: 120km (75 miles)
Armament: 1 × 37.2mm (1.46in) KwK 34(t) gun
Armour: 35mm (1.38in)

Panzer 38(t) Ausf C
This 1940-vintage Panzer 38(t) Model C sports two noticeable features typical of the design: the 37mm gun barrel set low in the rectangular mantlet and a large turret roof commander's cupola.

Panzerkampfwagen 38(t) Ausf C
Crew: 4
Production: 1939–1942
Weight: 10.5 tonnes (10.3 tons)
Dimensions: length: 4.61m (15ft 1in); width: 2.14m (7ft); height: 2.4m (7ft 10in)
Engine: 92kW (125hp) Praga EPA
Speed: 42km/h (26mph)
Range: 250km (155 miles)
Armament: 1 x 37mm (1.46in) KwK 38(t) L/47.8; 2 × 7.92mm (0.31in) ZB-53 MGs
Armour: 8–30mm (0.31–1.18in)

Division, but were all destroyed in these battles. Thereafter, the tank was employed only in rear area anti partisan duties.

Panzer 38(t) medium tank

After their March 1939 occupation of Czechoslovakia, the Germans took over the few LT-38 (Panzer 38(t)) light tanks already completed. In May–November 1939, BMM manufactured 150 Panzer 38(t) Model A tanks. These mounted either a 37.2mm (1.46in) Skoda or German 37mm (1.46in) gun and were protected by 8–25mm (0.3–1in) armoured plates. Subsequently, in 1940, BMM produced 325 slightly modified Model B, C and D vehicles. Next, between November 1940 and October 1941, BMM delivered 525 Model E and F vehicles, which sported an additional 25mm (1in) armour plate on the hull front. Finally, BMM constructed 321 Model G tanks in 1941–42, powered by an improved 112kW (150hp) engine; this increased the tank's operational range to 250km (155 miles). Of the 1414 Panzer 38(t)s completed, 72 were command tank variants.

During the September 1939 invasion of Poland, 80 Model A tanks served mainly in reconnaissance roles. Subsequently, in the May 1940 Western campaign, the Germans employed 228 Panzer 38(t)s out of an overall force of 2575 tanks. Some 754 Panzer 38(t) tanks participated in the June 1941 German invasion of the Soviet Union. Heavy losses ensured that just 522 Panzer 38(t)s remained operational in the East in April 1942. From summer 1942 onwards, by which time the 38(t) was obsolete, the Germans transferred the remaining vehicles to garrison duties or converted them into a range of armoured vehicles. By July 1944, 228 Panzer 38(t)s remained on occupation duties throughout Nazi controlled Europe.

Panzer III medium tank

The Panzer III medium tank was the first combat tank design the Germans developed. In 1936, Daimler-Benz completed 10 pre-production five-man Model A vehicles. These mounted the 3.7cm (1.46in) KwK L/45 gun, weighed 15 tonnes (14.8 tons) and were protected by 15mm (0.59in) armour. The vehicle's suspension featured five independently sprung medium-sized bogie-wheels with two return rollers. In 1937, factories produced 15 Model B tanks, with a suspension with eight small bogie-wheels per side grouped in pairs and suspended on two semi-elliptical springs, together with three return-rollers.

In 1937–38, Daimler-Benz produced 15 Model Cs with a suspension of eight bogie-wheels per side suspended from three semi-elliptic leaf-springs. Next, the company manufactured 40 Model Ds, which incorporated armour up to 30mm (1.18in) thick that raised the vehicle's weight to 19.3 tonnes (19 tons). Finally, some 96 Model E tanks were manufactured in 1939. The Model E featured a more powerful 239kW (320hp) engine and a suspension of six independent-bogied road-wheels

Panzer III Ausf F, 2nd Panzer Division

This Panzer III served with the 2nd Panzer Division during Operation 'Barbarossa'. The division fought as part of Army Group Centre, which was tasked with the failed push to capture Moscow.

Panzerkampfwagen III Ausf F

Crew: 5
Production: 1939–41
Weight: 21.8 tonnes (21.4 tons)
Dimensions: length: 5.38m (17ft 8in); width: 2.91m (9ft 6in); height: 2.44m (8ft)
Engine: 220kW (296hp) Maybach HL120TRM
Speed: 40km/h (25mph)
Range: 165km (102 miles)
Armament: 1 x 7.5cm (2.95in) KwK37 L/24 and 1 x 7.92mm (0.31in) MG13s
Armour: 10–50mm (0.39–1.96in)

Panzer III firepower				
Models	**A–F, early G**	**Late G–early J**	**Late J, L, M**	**N**
Main armament	37mm L/45	50mm L/42	50mm L/60	75mm L/24
Total produced	673	2815	1969	666

suspended from transverse torsion bars. All of these models saw service during the September 1939 invasion of Poland and the April 1940 occupation of Norway.

The Germans produced 435 Model Fs – the first general-production model – in 1939–40. This design proved highly reliable, thanks to the experimentation undertaken with the previous five designs. Combat experiences in the West in mid-1940, however, convinced the Germans that the Panzer III needed up-gunning and up-armouring. Consequently, the 450 Model Gs completed in 1940–41 mounted (with the exception of the first 90 vehicles) the more powerful 5cm (1.96in) KwK L/42 gun. Simultaneously, workshops retro-fitted existing Model A–F tanks with this gun. Next, in 1940–41, the Germans constructed 310 Model H tanks, which incorporated 30mm (1.18in) appliqué plates bolted onto the existing frontal armour. This increased the vehicle's weight to 21.6 tonnes (21.3 tons), which was compensated for by the introduction of wider tracks.

The first 250 Model J tanks mounted the 5cm (1.96in) L/42 gun, but featured

Panzer III, Operation 'Barbarossa', summer 1941

With equipment laden on its hull deck, this Panzer III tank, with panzergrenadiers sitting on top of it, moves towards a burning village during Operation 'Barbarossa', the summer 1941 Axis invasion of the Soviet Union.

Panzer III Ausf M

A side view of a Panzer III Model M, which is identifiable by its long-barrelled 5cm L/60 cannon and the curved additional *Schürzen* side-armour panels affixed to its turret sides.

Panzerkampfwagen III Ausf M

Crew: 5
Production: 1942–43
Crew: 5
Weight: 24 tonnes (23.6 tons)
Dimensions: length: 6.28m (20ft 7in); width: 2.95m (9ft 9in); height: 2.50m (8ft 2in)
Engine: 220kW (296hp) Maybach HL120TRM
Range: 155km (96 miles)
Speed: 40km/h (25mph)
Armament: 1 x 50mm (1.96in) KwK 39 L/60; 2 x 7.92mm (0.31in) MG
Armour: 10–50mm (0.39–1.96in)

integral (rather than bolted-on) 50mm (1.96in) thick frontal armour. The remaining 2266 Model J tanks mounted the more potent long-barrelled 5cm (1.96in) KwK 39 L/60 gun, which increased the vehicle's weight to 22.3 tonnes (21.9 tons). The Model L featured more effective spaced armour as well as a modified suspension to compensate for the new long 50mm (1.96in) gun, making the Model J nose-heavy. The Germans produced 703 Model L tanks in June–December 1942.

Subsequently, the production run of the Model M, which ran from October 1942 to February 1943, delivered 292 vehicles. The Model M sported novel thin armour side-skirts (*Schürzen*) to protect its turret and wheels from infantry-borne hollow-charge anti-tank weapons. In 1942–43, German factories delivered 666 Model N tanks, the final version of the Panzer III. This design was similar to its predecessor except that it mounted the short-barrelled (24-calibre) 75mm (2.95in) gun fitted in the early Panzer IV. This gun had a poor anti-tank performance, but was ideal for the heavy close-fire support role for which the Model N was intended. In total during 1936–43, German factories constructed 6123 Panzer III tanks. In addition, German firms produced 381 Panzer III command tank variants.

Just 98 Model A–F tanks fought in the September 1939 Polish campaign, although 349 Panzer IIIs participated in the May 1940 Western offensive. During 1941, the new 50mm (1.96in)-gunned Model G performed well in the summer

1941 German onslaught against the Soviets. Here, the Germans committed 1401 Panzer IIIs, of which 874 mounted the 50mm (1.96in) L/42 gun. In both the East and in North Africa between late 1941 and early 1943, the 50mm (1.96in) L/60-equipped Models J–M remained the stalwarts of the German panzer force. From late 1943, the Germans steadily withdrew this now-outclassed tank type from front-line service, although 80 Model M command tanks were operational in 1944. The remaining 704 Panzer III tanks served in garrison units across occupied Europe.

Panzer IV medium tank

The Panzer IV's developmental origins began in early 1934 when it was decided to design a medium close-support tank with a low-velocity, short-barrelled, 75mm (2.95in) gun to provide fire-support to the lighter Pz I, II and III tanks. Krupp produced 35 pre-production tanks in 1936, designated Panzer IV Model A. This design's superstructure overhung the hull sides, which facilitated subsequent up-gunning and ample internal high-explosive ammunition stowage for its 7.5cm (2.95in) KwK L/24 gun. The suspension consisted of four bogies per side, each of which carried two small rubber-tyred wheels supported by four return rollers. The tank was powered by a 186kW (250hp) engine, weighed 17.3 tonnes (17 tons) and had armour of up to 20mm (0.79in) thickness.

Panzerkampfwagen IV Ausf B
Crew: 5
Production: 1937–38
Weight: 20.7 tonnes (20.3 tons)
Dimensions: length: 5.92m (19ft 3in); width: 2.83m (9ft 3in); height: 2.68m (8ft 10in)
Engine: 220kW (296hp) Maybach HL120TR
Speed: 40km/h (25mph)
Range: 200km (124 miles)
Armament: 1 x 7.5cm (2.95in) KwK37 L/24; 1 x 7.92mm (0.31in) MG13s
Armour: 5–30mm (0.19–1.18in)

Panzer IV Ausf B, France 1940
The 'Bison' insignia identifies this vehicle as a Panzer IV Model B of the 10th Panzer Division during the 1940 Western Campaign; note the tall drum-shaped commander's cupola and eight small road wheels.

Panzer IV Ausf F1, 5th Panzer Division
The 'Red Devil' emblem identifies this up-armoured Panzer IV Model F1, equipped with the short-barrelled 7.5cm L/24 gun, as serving in the 5th Panzer Division, probably during the 1941 Axis invasion of the Soviet Union.

Panzerkampfwagen IV Ausf F1
Crew: 5
Production: 1937–38
Weight: 22.3 tonnes (24.6 short tons)
Dimensions: length: 5.92m (19ft 3in); width: 2.83m (9ft 3in); height: 2.68m (8ft 10in)
Engine: 220kW (296hp) Maybach HL120TR
Speed: 42km/h (26mph)
Range: 200km (124 miles)
Armament: 1 x 7.5cm (2.95in) KwK37 L/24; 1 x 7.92mm (0.31in) MG13s
Armour: 10–50mm (0.39–1.97in)

In 1937, the Germans produced 45 Model Bs that featured 30mm (1.18in)-thick armour, which increased the vehicle's weight to 17.7 tonnes (17.4 tons). A larger 237kW (320hp) engine powered the Model B to compensate for the vehicle's increased weight. In 1938–39, the Germans next delivered 140 Model C's, the first production model; this variant incorporated minor modifications to the armour. Next, in 1939–40, German factories produced 248 Model Ds that featured 20mm (0.79in)-thick rear/side armour; this again increased the tank's weight to 20 tonnes (19.6 tons).

Subsequently, the Germans produced 233 Model E tanks during 1940–41. This variant sported nose plate armour thickened to 50mm (1.96in), appliqué armour plates bolted to the sides, and a redesigned commander's cupola. Combat experience in Poland and France demonstrated that the Panzer IV was tactically sound.

In 1941, the Germans produced 975 up-armoured Model F1 vehicles, which sported 50mm (1.96in)-thick single-sheet frontal armour plates and 30mm (1.18in) side plates. These enhancements increased this variant's weight to 22.3 tonnes (21.9 tons); to offset this, the Germans fitted widened tracks that enabled the tank to maintain its maximum road speed at 42km/h (26mph). The modified F1 tank first saw service against the British in North Africa in 1941, while 548 Panzer IVs

Panzer IV Ausf H
This Panzer IV Model H sports five armoured side-skirts (the central one of which has become detached) to protect the tracks from man-held shaped-charge anti-tank weapons.

Panzerkampfwagen IV Ausf H
Crew: 5
Production: 1943–44
Weight: 27.6 tonnes (27.1 tons)
Dimensions: length: 7.02m (23ft); width: 2.88m (9ft 5in); height: 2.68m (8ft 10in)
Engine: 220kW (296hp) Maybach HL120TR
Speed: 38km/h (23mph)
Range: 210km (130 miles)
Armament: 1 x 7.5cm (2.95in) KwK40 L/48; 2 x 7.92mm (0.31in) MG34s
Armour: 10–80mm (0.39–3.14in)

participated in the June 1941 invasion of the Soviet Union. In total, 975 Model F1s were delivered during 1941.

In 1942, German firms delivered 1724 Model F2 and G tanks. In March 1942, the Model F2 appeared, which mounted the longer-barrelled 7.5cm (2.95in) KwK 40 L/43 gun, with its good anti-tank capabilities. This up-gunning reflected the German response to encountering the superior Soviet T-34 medium and KV-1 heavy tanks. Fitted with this potent gun, the Model F2 proved to be a match for the T-34 and superior to the British cruiser tanks encountered in North Africa. Mounting the longer gun, however, increased the tank's weight to 23.6 tonnes (23.2 tons) with a consequent reduction in its top speed to 40km/h (25mph). In August, the Model G, which sported an improved double baffle muzzle-brake on

Weight comparison of Panzer IV variants

Model	A	B	D	F1	F2	H
Weight (tonnes)	17.3	17.7	20	22.3	23.6	25
Weight (tons)	17	17.4	19.6	21.9	23.2	24.6

the 75mm (2.95in) gun, entered service. Model G tanks produced from October 1942 mounted an even longer-barrelled 75mm (2.95in) L/48 gun as well as armoured side-skirts to deflect hollow-charge anti-tank rounds.

The up-armoured Model H entered service in early 1943, sporting 80mm (3.14in)-thick hull nose armour. The design's increased weight of 25 tonnes (24.6 tons) reduced the vehicle's top road speed to 38km/h (23mph). German firms delivered 3073 Panzer IV tanks in 1943 and 3161 in 1944–45. The Nibelungenwerke factory in Austria delivered 2392 Model J vehicles in 1944–45. This variant sported a redesigned hull and removed the electrical turret traverse system, which enabled more fuel to be carried; the design therefore had an impressive 322-km (200-mile) by-road operational range. In addition, the Model J sported easier-to-produce mesh side-skirts rather than solid plate ones. In total, the Germans manufactured more than 7000 Panzer IV tanks between 1936 and 1945.

The Panzer IV, across its 10 models, served in every campaign the Germans fought during World War II. From 1939 to mid-1944, the Panzer IV generally matched its opponents in performance, with the main exception of combat in the East during the winter of 1941–42 before the up-gunned and up armoured Models F2 and G appeared. Even in 1944, the Panzer IV remained the mainstay of the German defence on all fronts. By mid-1944, however, even the Model J was becoming outclassed: its poorly sloped armour was vulnerable to the latest Allied anti-tank guns and the lethality of its gun had been eclipsed by newer, more potent, weapons.

Panthers by roadside in France, 1944

Frontal view of two Panther tanks, the front one a late-production Model D with redesigned 'rounded' (rather than drum-shaped) turret-roof cupola, and the rear vehicle a Model A or G with the new cupola and ball-mounted bow machine gun.

Panzer V Panther medium tank

The shock encounter with the potent Soviet T-34 medium and KV-1 heavy tanks during the summer 1941 Axis invasion of the Soviet Union led the Germans to develop more powerful combat tanks to counter these modern Soviet designs. This developmental work incorporated three features of the T-34 then absent in existing German tanks: sloped all-round armour for optimum shot deflection; large road wheels and wide tracks for speed and mobility; and a long-barrelled, over-hanging, large-calibre gun that produced high lethality. In January 1942, MAN and

Daimler Benz began designing rival 30-tonne (29.5-ton) VK3002 prototypes that incorporated these three features; in May, the former was selected for production. Powered by a 485kW (650hp) engine, the vehicle had sloped armour, interleaved wheels suspended from torsion bars, and a turret situated well back to mount a new long-barrelled 75mm (2.95in) L/70 gun.

Over-hasty design and production meant that the finished pre-production vehicles weighed 43 tonnes (42.3 tons), well above the target weight of 35 tonnes (34.4 tons). This excessive bulk caused numerous mechanical problems in the Panther – especially strain on the gearbox and transmission – that the Germans never entirely solved. To cope with these problems, the Germans installed a more powerful 522kW (700hp) Maybach engine and a more durable gearbox in the first production model, the Panzer V Panther Model D. These modifications only partially ameliorated the numerous mechanical problems that dogged early Panther

Panzer V Ausf D, winter camouflage

Finished in white-washed winter camouflage, this early Model D Panther (identifiable by the drum-shaped turret-roof cupola and by the lack of a bow machine gun) is designated II01 – the second battalion commander's personal vehicle.

Panzerkampfwagen V Ausf D

Crew: 5
Production: 1943–44
Weight: 47.4 tonnes (46.7 tons)
Dimensions: length: 8.86m (29ft); width: 3.42m (11ft 2in); height: 2.95m (9ft 8in)
Engine: 522kW (700hp) Maybach HL230P30
Speed: 46km/h (29mph)
Range: 200km (124 miles)
Armament: 1 x 7.5cm (2.95in) KwK42 L/70; 2 x 7.92mm (0.31in) MG34
Armour: 16–110mm (0.62–4.3in)

tanks. The five-man Panther Model D possessed 80–110mm (3.2–4.3in)-thick frontal armour. The Germans produced 600 Model D tanks between November 1942 and September 1943. Despite these mechanical problems, 250 Model D tanks participated in Operation 'Citadel', the 5 July 1943 German offensive designed to encircle the Soviet forces deployed in the Kursk salient. By the second day of 'Citadel', however, just 21 per cent of the Panthers committed remained operational; many succumbed to mechanical failure, engine fires or enemy mines.

In the wake of 'Citadel', the Germans concluded that, as presently configured, the Model D was unfit for front-line service. As a result, the Germans implemented a series of modifications to late-production vehicles, including a new cast cupola that improved the commander's protection and field of vision. The Germans also introduced an improved Panther Model A vehicle in September 1943. This vehicle was distinguishable from its predecessor by its ball mounted bow machine gun, which enabled a wider field of fire. Other design modifications made on the Model A included strengthened road wheels to counteract the overloading of the suspension. Minor alterations to the tank's transmission and gearbox during the autumn progressively corrected the earlier mechanical problems. At the same time, introduction of additional cooling pipes and abandonment of

Panther Ausf A, LSSAH Division
Sporting an unusual three-tone summer disruptive pattern camouflage scheme, the three aerials evident on this Model A Panther identifies it as a command tank variant.

Panzerkampfwagen V Ausf A
Crew: 5
Production: 1943–44
Weight: 49.4 tonnes (48.6 tons)
Dimensions: length: 8.86m (29ft); width: 3.42m (11ft 2in); height: 2.98m (9ft 10in)
Engine: 522kW (700hp) Maybach HL230P30
Speed: 46km/h (29mph)
Range: 200km (124 miles)
Armament: 1 x 7.5cm (2.95in) KwK42 L/70; 2 x 7.92mm (0.31in) MG34
Armour: 16–110mm (0.62–4.3in)

Panther variants compared				
	Weight	Number Produced	Commander's turret cupola	Maximum road speed
Model D	43 tonnes (42.3 tons)	842	Drum	55km/h (34mph)
Model A	44 tonnes (43.3 tonnes)	2200	Cast	46km/h (29mph)
Model G	44.8 tonnes (44.1 tons)	4010	Cast	46km/h (29mph)

watertight sealing reduced the Panther's susceptibility to engine fires, the weakness that had crippled so many vehicles at Kursk.

From October 1943, the Germans fitted all new production Panthers with Zimmerit anti-magnetic mine paste applied to the hull and turret; this was designed to hinder the attachment of hollow charge magnetic anti-tank mines. The paste gave the vehicle a distinctive uneven cement like texture, though the application was abandoned in September 1944 as an unnecessary luxury since Germany's opponents rarely used magnetic mines. In total, German tank firms produced 1768 Panther Model A tanks in 1943–44.

The final Panther variant, the Model G, entered service in February 1944. The design sported a better-sloped redesigned hull, thus enhancing the vehicle's survivability. German factories manufactured 3740 Panther Model G tanks in 1944 and a further 270 in early 1945; some five per cent of these were constructed as command variants. The Model G Panther was a stalwart of the desperate defensive stands the German Army undertook in the last year of the war. One exceptional example should suffice to convey the potency of the Model G's gun. On 8–29 July, during bitter defensive stands against the Allied D-Day invasion, the Model G tank of SS NCO Ernst Barkmann, the commander of the 4th Company, of the 2nd SS Panzer Division *Das Reich*, alone destroyed 18 Allied Sherman tanks. By 1944, the Panther had become a highly effective battle tank by the standards of World War II. It ably combined firepower, protection and mobility into a first-rate fighting machine that remained generally superior to its opponents until the war's end.

Panzer VI Tiger heavy tank

The development of the Panzer VI Tiger heavy tank originated in Germany's summer 1941 combat experiences against the Red Army's potent new T-34 medium and KV heavy tanks. These encounters prompted the Germans to develop both a new medium tank (the Panther) and a new heavy tank design that eventually became the Panzer VI Model E Tiger. In mid-1942, initial development of this heavy tank centred respectively on the rival Henschel- and Porsche-designed VK4501(H) and VK4501(P) prototypes. These experimental tanks utilized a modified version of the existing VK3601 chassis and mounted the 8.8cm (3.4in) tank gun version of the Famous 'Flak 88'. In August 1942, the VK4501(H) entered general production at Henschel's Kassel factory as the Panzer VI Model E Tiger. During a 23-month-long production run that ended in June 1944, Henschel completed 1354 Tigers.

Visually, the five-man Tiger was a squat and angular heavy tank, not dissimilar to the smaller Panzer IV. The vehicle mounted the powerful 56 calibres-long 8.8cm (3.4in) KwK 36 gun and weighed a massive 56 tonnes (55.1 tons). The tank's high lethality and survivability performance accounted for much of this weight; the tank was protected, for example, by 100mm (3.94in)-thick vertical frontal armour and 80mm (3.14in) side/rear armour.

The first 250 Tigers manufactured featured a 479kW (642hp) engine, while subsequent vehicles employed the more powerful 522kW (700hp) HL 230 powerplant. Tigers equipped with the latter engine could achieve a top road speed of 38km/h (24 mph) and a cross-country speed of 20km/h (12mph).

As the Tiger was so large, Henschel had to design a clever double-track system. During operational service, the Tiger utilized 725mm (28.5in)-wide battle tracks that helped keep its ground-pressure ratio to reasonable levels. When being transported by railway flatcars, however, these battle tracks were replaced with narrower

Tiger in Normandy
Heavily camouflaged with foliage, a Tiger I Ausf E from the schwere SS-Panzer-Abteilung 102 advances down a road somewhere in France, heading for the front line.

Panzerkampfwagen VI Ausf E
Crew: 5
Production: 1942–45
Weight: 50.8 tonnes (49.9 tons)
Dimensions: length: 8.45m (27ft 8in); height: 3m (9ft 10in); width: 3.56m (11ft 8in)
Engine: 522kW (700hp) Maybach HL230 V12 petrol engine
Speed: 38km/h (24mph)
Range: 140 km/h (87mph)
Armament: Main gun: 88mm (3.4in) KwK36 L/56 gun; 2 x 7.92mm (0.31in) machine guns
Armour: 25–120mm (1–4.72in)

Tiger I Ausf E
Prominent in this side view of a Tiger tank are the eight large interleaved road wheels, plus driver and idler wheels, around which formed its running gear.

520mm (20.4in) transport tracks after the outer layer of road wheels had been removed. Overall, although the Tiger was a lethal and highly survivable heavy tank, its combat effectiveness was hampered by its sheer size, substantial weight, poor cross-country tactical mobility, mechanical unreliability, and heavy fuel consumption.

The first four Tigers made their operational debut on 29 August 1942 at Mga near Leningrad, where unsuitable terrain limited their impact. However, the tank subsequently demonstrated its potent lethality; on 29 December 1942, five Tigers of the 502nd Battalion knocked out 12 Soviet T-34 medium and T-60 light tanks near Leningrad.

Comparison of Tiger and King Tiger

	Service	Weight	Number produced	Main armament
Tiger	1942–45	56 tonnes (55.1 tons)	1354	8.8cm KwK 36 L/56
King Tiger	1944–45	69.4 tonnes (68.3 tons)	489	8.8cm KwK 43 L/71

That same month, Tiger tanks engaged Western Allied forces in Tunisia for the first time.

During 1943–44, meanwhile, the Tiger tank also built up a formidable record on the Eastern Front. By late-1944, however, the Tiger was becoming out-classed by the heavily protected and lethally armed new Soviet Josef Stalin (JS) series heavy tanks, which embraced well-sloped armour rather than the Tiger's mostly vertical plates. Given that Tiger tank production ceased in June 1944 (in favour of the King Tiger), and with severe losses, the Tiger had already become a very rare vehicle by late 1944. The number of Tiger tanks the Germans fielded peaked at 631 in June 1944, but this had dwindled to 243 by December.

Panzer VI B King Tiger

The Panzer VI Model B Tiger II ('King Tiger') heavy tank was a logical development of the Tiger that incorporated the Panther tank's well-sloped armour. The King Tiger remained a very rare vehicle: only 489 saw service during the war's last 14 months. From January 1944 onwards, Henschel's Kassel factory was contracted to produce 1500 King Tigers. As a result of five devastating Allied air attacks, however, King Tiger production consistently lagged behind the planned schedule of 659 tanks delivered by 31 March 1945. Some 23 of the 489 tanks constructed were command variants fitted with either a powerful 20- or 30-Watt radio transmitter. Thanks to its combination of lethal firepower and virtual invulnerability to Allied fire, the vehicle was one of the most formidable tanks of the entire war.

The King Tiger's developmental origins began in 1943, as a logical bid to up-gun and up-armour the rather dated Tiger I design. The tank mounted the extremely potent long-barrelled 8.8cm (3.4in) KwK 43/3 L/71 gun. Weighing 69.4 tonnes (68.3 tons) and powered by a 522kW (700hp) engine, the King Tiger had excellent survivability, with a maximum of 185mm (7.28in) thick well-sloped armour on its turret face. The first 50 production vehicles built were unique in sporting a round-fronted Porsche turret rather than the standard Henschel flat-fronted one. Although a potent fighting vehicle, the King Tiger was a fuel-consumptive, cumbersome and complexly engineered tank that was plagued by mechanical problems, particularly with its drive system.

Always a rare vehicle in German service – the number in service peaked at just 219 vehicles in February 1945 – the Tiger II served in nine Wehrmacht and three SS heavy tank battalions, plus two independent companies. Some

TIGERS AT VILLERS-BOCAGE

Allied combat against the lethal and superbly protected Tiger tank in the 1944 Normandy campaign highlighted that, in a more static defensive role (where its mechanical unreliability was less significant), the tank was a formidable weapon. On 13 June 1944 at Villers-Bocage, for example, a handful of Tigers, including that of SS panzer ace Michael Wittmann, flung back and severely mauled an entire British brigade, destroying 47 vehicles and inflicting 257 personnel casualties.

Similarly, on 8 August 1944 at Vire, 10 Tigers of the 1st Company, 102nd SS Heavy Tank Battalion, destroyed 24 Allied tanks.

53 King Tigers fought during the summer 1944 Normandy campaign, where most were lost. A further 52 Tiger IIs – one-third of the total then in service – also played a key part in the December 1944 German Ardennes counter-offensive. The 501st SS Heavy Tank Battalion's King Tigers formed part of Battle-Group Peiper (BGP), which was to rapidly exploit any success towards Antwerp before the Allies could react. Because the terrain over which BGP was to advance was so difficult, the lumbering King Tigers advanced at the rear of the column. Only on 20 December, when BGP's advance had stalled at Stoumont, did 10 King Tiger tanks catch up with the main force. Subsequently, Allied counter-blows surrounded BGP at La Gleize; having run out of fuel and ammunition, the

Panzerkampfwagen Tiger Ausf. B
Crew: 5
Production: 1942–44
Dimensions: length: 10.26m (33ft 8in); height: 3.09m (10ft 1in); width: 3.75m (12ft 3in)
Weight: 69.4 tonnes (68.3 tons)
Engine: 522kW (700hp) Maybach HL230P30 V12 petrol engine
Speed: 37km/h (23mph)
Range: 140km (87mph)
Armament: Main gun: 88mm (3.4in) KwK43 L/71 gun; 2 x 7.92mm (0.31in) machine gun
Armour: 25–185mm (1–7.28in)

'King Tiger'
The length and height of the King Tiger's turret stands out in this left-hand side-view of the tank; this was necessary to house the large breech and recoil of the powerful 8.8cm KwK 43 L/71 cannon.

Tiger battalion
A King Tiger battalion assembled on a proving ground in Germany. The turret of the foreground vehicle clearly shows the rippled surface of Zimmerit anti-magnetic mine paste.

disabled its 35 tanks – including six King Tigers – and fought its way out on foot. During the Ardennes counter-offensive, the Germans lost 20 King Tigers, some 40 per cent of the force committed. Just five were lost to enemy fire, while the remainder were abandoned due to lack of fuel, mobility damage or crashes.

The last King Tiger actions in the West occurred in April 1945, when the 510th and 511th Battalions' 13 tanks – the last ever produced – were lost during desperate defensive actions fought to stem the headlong Allied advance across the Reich.

From October 1944 onwards, the King Tiger also fought on the Eastern Front after the arrival of the 501st Battalion's tanks. Subsequently, in mid-January 1945, the 503rd SS Heavy Tank Battalion, after being issued with 39 newly produced King Tigers, was rushed to central Poland to help shore up the disintegrating front in the face of the Soviet Vistula-Oder offensive. Desperate defensive fighting on the borders of the Reich reduced this unit to just two operational tanks by 20 March 1945. By the war's end, just nine King Tigers remained on the Eastern Front.

SELF-PROPELLED ANTI-TANK GUNS, ASSAULT GUNS AND TANK DESTROYERS

In addition to tanks, German self-propelled anti-tank guns (SPATGs), assault guns and tank destroyers represented a second group of mobile direct-fire armoured vehicles that played a key part in the offensive successes and dogged defensive last-stands achieved by the Wehrmacht during the war. These vehicles existed with the primary purpose of destroying enemy armoured vehicles, particularly tanks. German self-propelled anti-tank guns and tank destroyers both sported a powerful anti-tank gun in a fixed superstructure mounted on top of the chassis of an armoured vehicle or tank. German self-propelled anti-tank guns were typically improvised vehicles that utilized the chassis of obsolete tanks. Tank destroyers typically were deliberately designed variants of a German combat tank. Assault guns were similar vehicles that started the war with a focus on high-explosive fire, but increasingly had their anti-tank capability upgraded so that they became cheaper to make alternatives to combat tanks.

This chapter examines the nine most important examples of these vehicles.

Panzerjäger II für 7.62cm (3in) Pak 36(r) oder 7.5cm (2.95in) Pak 40 Marder II SPATG

During the winter of 1941–42, the Germans realized they urgently needed mobile anti-tank assets to counter the formidable new Soviet T-34 and KV-1 tanks. In 1942, Alkett produced 235 improvised Marder II (Sdkfz 132) SPATGs. This married the chassis of a Panzer II Model D–F tank to the potent Soviet 76.2mm (3in) M36 field gun, which the Germans had captured in large numbers.

7.5cm (2.95in) Panzerjäger Marder II
Crew: 4
Production: 1942–43
Weight: 11.76 tonnes (11.57 tons)
Dimensions: length: 5.85m (19ft 2in); width: 2.16m (7ft); height: 2.5m (8ft 2in)
Engine: 103kW (140hp) Praga EPA or EPA/2
Speed: 42km/h (26mph)
Range: 185km (115 miles)
Armament: 1 x 75mm (2.95in) Pak 40 gun
Armour: 14.5–35mm (0.57–1.38in)

SdKfz 132 Panzerjäger Marder II
The expedient nature of the Marder II (Sdkfz 132) is evident in this view; the vehicle has a very high profile and the crew obtain minimal protection from the small three-sided gun shield.

On top of the chassis, Alkett mounted a high, box-like, 14.5mm (0.57in)-thick steeply sloping armoured superstructure. The gun was mounted on top of this within a three sided open 10mm (0.39in)-thick gun-shield. This vehicle could only have modest protection because, with the larger gun, it weighed 10.7 tonnes (10.5 tons) – close to the chassis' maximum load.

The Germans converted a further 531 (Sdkfz 131) Marder IIs in 1942–43. Mounted on Model A–C chassis, this variant mounted the gun directly onto the chassis without any armoured superstructure. The gun sat within a single section, lightly armoured, three-sided shield fitted directly onto the chassis, leaving the fighting compartment exposed at the top and rear.

Once the stocks of captured Soviet 1936 field guns had been utilized during 1942–44, the Germans also produced 1217 Sdkfz 131 Marder II vehicles that mounted the German 7.5cm (2.95in) Pak 40/2 L/46 gun onto any available Panzer II chassis. This Marder II variant mounted the entire gun with its original shield, recoil system, and traversing and elevating gears on a platform positioned high on the hull; the gun sported a large, almost rectangular, protective shield 10mm (0.39in) thick. In 1942–44, these Marder IIs served primarily in infantry division anti-tank battalions in the East, where their potent anti-tank capabilities were much valued.

Panzerjäger 38(t) für 7.62cm (0.3in) Pak 36(r)

During 1942, the Germans developed a new improvised SPATG, the Marder III. This married the potent captured Soviet 76.2mm (3in) M36 gun (rechambered to take the standard German 7.5cm/2.95in Pak round) to the Panzer 38(t) Model G tank chassis. Factories fitted an armoured superstructure that comprised a small, three-sided, 10mm (0.39in)-thick armoured shell; the entire gun and carriage was fixed to the chassis' top by a purpose-built turntable. The vehicle

Sdkfz 131 Marder II

This 7.5cm (2.95in)-gunned Marder II (Sdkfz 131) sports an MG 34 close-defence machine gun mounted on the top of one side of the open-topped superstructure; this was a common in-theatre field modification.

Panzerjäger 38(t) 7.5cm PaK 40 Ausf M Marder III
With its rear-located fighting compartment, this Marder III, with desert camouflage, sports a barrel rest that supported the gun when the vehicle was moving non-tactically over significant distances.

Panzerjäger 38(t) mit 7.5cm (2.95in) PaK40/3 Ausf M
Crew: 4
Production: 1943–44
Weight: 11.6 tonnes (11.4 tons)
Dimensions: length: 4.95m (16ft 2in); width: 2.15m (7ft); height: 2.48m (8ft 1in)
Engine: 118kW (158hp) Praga C
Speed: 42km/h (26mph)
Range: 190km (118 miles)
Armament: 1 x 7.5cm (2.95in) PaK 40/3 L/46 gun
Armour: 14.5–35mm (0.57–1.38in)

also sported a hull front-mounted 7.92mm (0.31in) MG 37(t) machine gun for close-range defence. The heavy 76.2mm (3in) gun increased the vehicle's weight to 10.7 tonnes (10.5 tons), close to the chassis' maximum capacity, which limited the protection that could be given to the crew. Some 344 76.2mm (3in)-equipped Marder IIIs were produced in 1942.

In 1942–43, the Germans also manufactured 418 Marder III vehicles that instead sported the powerful 7.5cm (2.95in) Pak 40/3 L/46 gun mounted on the Panzer 38(t) Model H chassis. This variant featured a redesigned superstructure with a longer and larger shield that provided better crew protection. Both these designs, however, had rear mounted engines that necessitated forward placement of the fighting compartment, which left the vehicles front-heavy. Late 75mm (2.95in)-gunned Marder IIIs (designated Model M) utilized a fundamentally redesigned chassis with the engine in the centre and a rear-located fighting compartment, creating a more stable fighting platform. German factories constructed 975 Model M vehicles in 1943–44.

Sturmgeschütz III assault gun

In the mid-1930s, the German artillery arm requested the development of a low-silhouetted non-turreted close infantry support AFV, or assault gun (Sturmgeschütz) with dual high-

StuG III lethality			
StuG III model:	**A–E**	**F**	**G**
Main armament	7.5cm KwK L/24	7.5cm StuK 40 L/43	7.5cm StuK 40 L/48
Muzzle velocity	420m/s (1378ft/s)	740m/s (2400ft/s)	770m/s (2525ft/s)

explosive and anti-armour capability. The four-man StuG III Model A appeared in 1940, when 184 were produced. It mounted the Panzer IV's short-barrelled 7.5cm (2.95in) KwK L/24 gun with limited traverse in a fixed armoured superstructure installed directly onto the Panzer III tank chassis. The vehicle carried 55 high-explosive and 21 armour-piercing rounds, plus eight smoke charges. The weight-saving absence of a turret enabled the vehicle to possess better protection than that of the Panzer III, with 50mm (1.96in)-thick frontal and 43mm (1.7in) side armour. In 1941, the Germans produced 548 slightly modified StuG III Models C, D and E vehicles, which incorporated a new six-speed synchro-mesh gearbox.

After encountering the formidable Soviet T-34 and KV-1 tanks in late 1941, the Germans desperately needed to increase their mobile anti-tank assets. Consequently, the significantly modified StuG III Model F entered production in May 1942, with 120 being built. This vehicle mounted the

StuG III Ausf D

The designation '112' identifies this Model D StuG III as the second vehicle in the first platoon of the 1st Company; note the spare tyre housed on the superstructure's rear roof.

Sturmgeschütz III Ausf D
Crew: 4
Production: 1940–45
Weight: 19.6 tonnes (19.2 tons)
Dimensions: length: 5.38m (17ft 8in); width: 2.92m (9ft 7in); height: 1.95m (6ft 5in)
Engine: 224kW (300hp) Maybach HL120TR
Speed: 40km/h (25mph)
Range: 160km (99 miles)
Armament: 1 x 7.5cm (2.95in) StuK 37 L/24 gun
Armour: 16–80mm (0.62–3.14in)

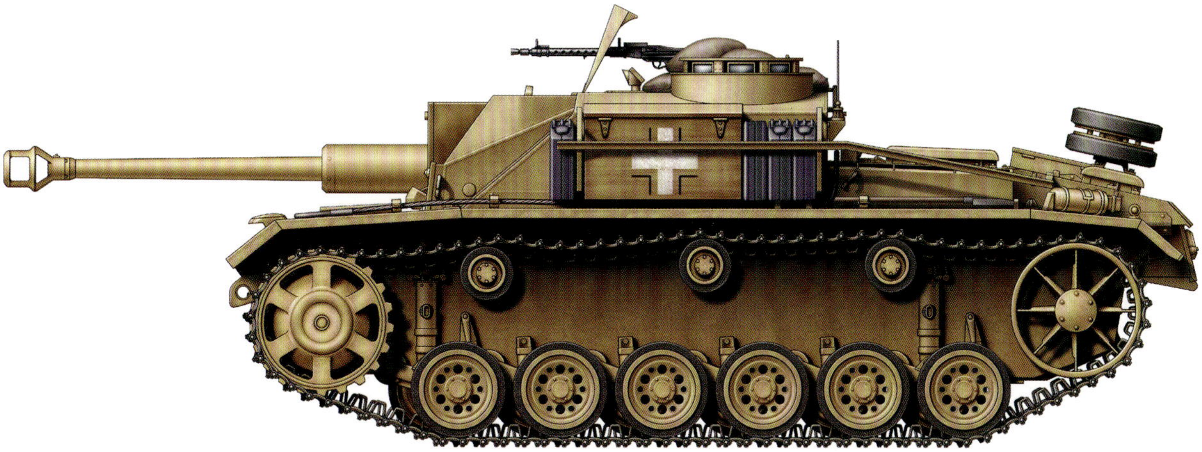

StuG 40 Ausf G
View of an early example of the final StuG III variant, the StuG 40 Model G, sporting a large rectangular mantlet; later Model Gs instead featured the 'Pig's Head' mantlet.

7.5cm (2.95in) Sturmgeschütz 40 Ausf G
Crew: 4
Production: 1940–45
Weight: 24 tonnes (23.6 tons)
Dimensions: length: 6.77m (22ft 2in); width: 2.95m (9ft 8in); height: 2.16m (7ft)
Engine: 224kW (300hp) Maybach HL120TR
Speed: 40km/h (25mph)
Range: 155km (96 miles)
Armament: 1 x 7.5cm (2.95in) StuK 40 L/48
Armour: 16–80mm (0.62–3.14in)

much longer-barrelled (43 calibres long) 7.5cm (2.95in) StuK 40 L/43 cannon as its main armament. This gave it a much-increased anti-armour capability, but also raised its weight to 21.6 tonnes (21.3 tons).

The next variant, the Model G, sported the more potent 7.5cm (2.95in) StuK 40 L/48 gun and thicker armour, which again raised the vehicle's weight to 24 tonnes (23.6 tons). The standard variant, the Model G, was mass-produced; 3041 vehicles were completed during 1943, 4851 in 1944, and just 123 in early 1945. The StuG III became a workhorse of the Germany Army, serving in all theatres and several roles. It made up for tank shortages in some panzer divisions; it equipped numerous independent assault gun formations, and it also served in some infantry division anti-tank units.

Jagdpanzer IV tank destroyer

In late 1943, the Germans developed the tank destroyer variant of the Panzer IV tank, the Jagdpanzer IV ('Hunting Tank IV'). This became the German Army's first deliberately designed tank hunting vehicle: hence the then-novel designation Jagdpanzer ('Hunting Tank'). Production of the latter commenced in January 1944. In August, all Panzer IV tank production ceased and these assets were redirected to produce this tank destroyer variant.

Between January and November 1944, the German armament firm of Vomag produced some 769 Jagdpanzer IV vehicles. The original development work had intended to mount the Panther tank's potent 7.5cm (2.95in) L/70 gun in the Jagdpanzer IV, but this intention fell foul of technical problems. Rather than delay the entire project, the German Army contracted Vomag to produce the Jagdpanzer IV utilizing the then-extant stocks of the proven 75mm (2.95in) KwK 40 L/48 gun as already mounted in the StuG III assault gun and Panzer IV combat tank. Unusually, this cannon was not mounted in the centre of the sloping front superstructure plate, but was offset 20cm (8in) to the right of the centre line. Early vehicles also possessed a muzzle brake on the main armament. However, as the vehicle had such a low silhouette – when horizontal the barrel was only 1.4m (4ft 7in) above the ground – this muzzle brake threw up dust from the deflected blast when the gun fired. This heavy back-blast obstructed the crew's vision and could give away the vehicle's position to the enemy. As a result, later vehicles had the muzzle brake deleted.

The Jagdpanzer IV vehicle possessed an excellent level of protection in large part thanks to its combination of a low silhouette and well-sloped armour, which maximized shot deflection. Its 60mm (2.4in)-thick upper and lower frontal nose plates, for example, were sloped at 45 and 57 degrees, respectively. The vehicle carried 30mm (1.18in)-thick plate on its superstructure sides. The vehicle weighed a modest (by the standards of 1944) 24.1 tonnes (23.7 tons) and was powered by a 224kW (300hp) Maybach engine. In 1944, the Panzer IV/70 gradually replaced the Marder II and III vehicles fielded by the tank destroyer battalions of German

Jagpanzer IV in Normandy
The extent to which the Germans used fieldcraft to conceal their tanks from Allied tactical air power during the 1944 Normandy campaign is evident in this view of a foliage-laden short-barrelled Jagdpanzer IV.

Jagdpanzer IV
The early interim Jagdpanzer IV was easily identifiable by its shorter 7.5cm KwK 40 L/48 gun and the gentle downward slope of its armoured superstructure's rear roof.

Jagdpanzer IV
Crew: 4
Production: 1943–45
Weight: Up to 27.6 tonnes (27.1 tons)
Dimensions: length: 6.85m (22ft 6in); width: 6.7m (22ft); height: 1.85m (6ft)
Engine: 224kW (300hp) Maybach HL120TRM
Speed: 38km/h (24mph)
Range: 210km (130 miles)
Armament: 7.5cm (2.95in) Pak 39 L/48 gun anti-tank gun; 1 x 7.92mm (0.31in) MG
Armour: 10–80mm (0.39¬–3.14in)

panzer divisions, but due to the small numbers produced it remained an uncommon vehicle.

Jagdpanzer IV/70 (Panzer IV/70) tank destroyer

In August 1944, once the technical challenges of mounting the modified 7.5cm (2.95in) StuK 42 L/70 cannon in the Jagdpanzer IV's low superstructure had finally been solved, production commenced of this new vehicle, known either as the Jagdpanzer IV/70, the Panzer IV/70 or the Panzer IV/70(V) tank destroyer.

The German Vomag factory produced 930 Panzer IV/70s between August 1944 and March 1945. Mounting the heavier and longer L/70 cannon required some alterations to the Jagdpanzer IV's original design. These modifications compensated for the increased weight exerted on the vehicle's nose by the sheer length of the L/70 gun, which projected 2.58m (8ft 6in) beyond the vehicle's front face. These alterations included, for added strength, redesigned interlocking upper and lower hull nose plates, and superstructure front and side plates. Mounting the L/70 gun in the Jagdpanzer IV's shallow armoured superstructure required that the gun's buffer and recuperator mechanisms be relocated above the barrel. Like its predecessor, the cannon was located offset, 20cm (8in) inches to the right of the centre line. The Panzer IV/70 was powered by the same 224kW (300hp) Maybach engine as its predecessor, located at the rear. Both the longer gun and the heavier armour increased the vehicle's weight to 25.8 tonnes (25.4 tons).

Comparison of Jagdpanzer IV and Panzer IV/70 production runs			
Model	**Jagdpanzer IV**	**Panzer IV/70**	**Panzer IV/70(A)**
Production run	Jan–Nov 1944	Aug 1944–March 1945	Aug 1944–March 1945
Number produced	769	930	278
Monthly average	70	116	35

By summer 1944, when Panzer IV/70 production finally commenced, the Germans were desperate to get as many AFVs to the front line as quickly as possible, yet Panzer IV/70 manufacture was already bedevilled with bottlenecks. Thus the Germans hastily redesigned a quicker to produce stopgap variant of the tank destroyer. This mounted the 7.5cm (2.95in) StuK 42 L/70 gun in a modified Panzer IV/70 superstructure placed directly on top of the unmodified chassis of the Panzer IV Model J tank; this was in mass production and thus immediately available in large numbers. This vehicle was designated Panzer IV/70 Zwischenlösung ('Interim') or IV/70(A) – the 'A' standing for the firm of Alkett, who designed the modified superstructure.

The Nibelungenwerke produced 278 Panzer IV/70(A)s between August 1944 and March 1945. The interim Panzer IV/70(A) looked different to the Panzer IV/70; it had a cutback rear superstructure and a vertical rear plate. The Panzer IV/70(A) also possessed enhanced armour protection of up to 120mm (4.72in) thickness, which

Jagpanzer IV, Ardennes Offensive
A Jagdpanzer IV, with German grenadiers lying on its roof, moves down a slope during the December 1944 Ardennes counter-offensive; this image suggests how easily the longer-barrelled Jagdpanzer IV/70 could get its gun barrel stuck in the ground when descending slopes.

Jagdpanzer IV, 21st SS Panzer Division

Aside from some minor modifications to its running gear, what made the Jagdpanzer IV/70 immediately identifiable from the Jagdpanzer IV was the significantly longer-barrelled 7.5cm (2.95in) L/70 cannon.

Jagdpanzer IV/70
Crew: 4
Production: 1944–45
Weight: 25.8 tonnes (25.4 tons)
Dimensions: length: 8.50m (27ft 10in); width: 3.17m (10ft 5in); height: 2.85m (9ft 4in)
Engine: 224kW (300hp) Maybach HL120TRM
Speed: 35km/h (22mph)
Range: 210km (130 miles)
Armament: 7.5cm (2.95in) Pak 42 L/70 anti-tank gun; 1 x 7.92mm (0.31in) MG
Armour: 10–80mm (0.39–3.14in)

increased its weight to 28 tonnes (27.6 tons). Thanks to its combination of lethality, survivability and mobility, the Panzer IV/70 was an effective AFV. However, it was produced in too small numbers to exert much impact on the inexorable Allied advance into the heart of the German Reich in 1944–45.

Jagdpanzer V Jagdpanther tank destroyer

In 1943–44, the Germans developed the Jagdpanzer V Jagdpanther ('Hunting Panther'), the tank destroyer variant of the Panther tank. The firm of MIAG, later assisted by MNH, manufactured 382 Jagdpanthers between February and April 1945, well short of the target figure of 1100. This vehicle comprised a standard Panther Model G tank chassis fitted with a very well-sloped armoured superstructure that housed the main gun – the potent 8.8cm (3.4in) Pak 43/3 cannon. The tank destroyer was protected by 80mm (3.14in)-thick frontal plates and 40–50mm (1.6–1.96in) armour on the sides and rear. This armour provided only modest survivability by the standards of 1944; this was augmented by the vehicle's low silhouette and the pronounced slope of most of these plates. The long sweeping lines of the Jagdpanther made it visually one of the most elegant AFVs of the war.

Despite its substantial weight of 45.5 tonnes (44.8 tons), the Jagdpanther still possessed good mobility, thanks to its powerful 522kW (700hp) Maybach engine, inter-leaved wheel suspension arrangement and wide tracks. These features enabled the vehicle to achieve an admirable 45km/h (28mph) maximum road speed, and a top cross-country performance of 24km/h (15mph). Given its short production

JAGDPANTHERS IN OPERATION 'NORTHWIND'

The eight operational Jagdpanthers of the 654th Army Battalion similarly played an important role in Operation 'Northwind', the January 1945 German offensive in Alsace-Lorraine. This attack involved a six-division thrust in the north that advanced southward to Strasbourg in order to meet a southern German pincer launched from the Colmar Pocket, the German-held salient jutting beyond the River Rhine onto French soil. The battalion's Jagdpanthers played a key role in the partly successful advance by the southern German pincer, although four were lost in the process.

programme and the limited number of vehicles completed, the Jagdpanther underwent no significant design modifications during its manufacturing run. This design achieved a good balance between potent firepower, impressive survivability and good mobility, making the Jagdpanther one of the most effective German AFVs of the entire war.

The Jagdpanther first entered combat in the later stages of the Normandy campaign. On 30 July, the 14 operational vehicles of the 2nd Company, 654th Heavy Anti-tank Battalion, played a crucial role in halting Operation 'Bluecoat'; this British offensive aimed to widen the frontage of the breakthrough already achieved to the west by the Americans at Avranches.

Jagdpanther
This left side view ably shows the Jagdpanther's sleek, low-silhouetted and well-sloped profile, as well as the extent of its barrel overhang.

Jagdpanther
Crew: 5
Production: 1944–45
Weight: 45.5 tonnes (44.8 tons)
Dimensions: length: 9.9m (32ft 6in); width: 3.43m (11ft 2in); height: 2.72m (8ft 10in)
Engine: 522kW (700hp) Maybach HL230P30
Speed: 45km/h (28mph)
Range: 160km (99 miles)
Armament: 1 × 8.8cm (3.4in) Pak 43/3 or 43/4 L/71 main gun; 1 × 7.92mm (0.31in) MG 34
Armour: 40–100mm (1.57–3.94in)

That day, these Jagdpanthers knocked out 16 Churchill tanks.

The Jagdpanther was typically deployed in small numbers, but in the Battle of the Bulge (the German December 1944 Ardennes counter-offensive) Jagdpanthers operated in greater mass. For this operation, 51 Jagdpanthers – 13 per cent of the total number ever manufactured – were committed. On 20 December 1944 at Dom Butgenbach, for example, the eight Jagdpanthers of the 560th Army Heavy Anti-tank Battalion provided intense fire support for the desperate assaults of the grenadiers of the 12th SS Panzer Division *Hitlerjugend*.

The minor offensive and defensive successes achieved by the Jagdpanther, however, made no discernible strategic impact on the war, given the small number of vehicles produced.

Jagdpanzer VI Jagdtiger heavy tank destroyer

During 1943–44, the Germans developed the Jagdpanzer VI Jagdtiger ('Hunting Tiger'), the tank destroyer variant of the famous King Tiger tank. In early 1944, the Army contracted the Austrian armaments firm of Steyr Daimler Puch in St Valentin to produce 150 Jagdtigers; in reality, the firm completed only 77 vehicles before the end of the war in May 1945.

The vehicle's chassis was a slightly extended version of that of the King Tiger tank. In the Jagdtiger, however, the King Tiger's slightly inwardly sloping hull sides were extended upwards. When married to a front and back plate, this created the rectangular superstructure fighting compartment, which housed its main armament, the potent 12.8cm (5in) Pak 44 L/55 cannon; this was the largest-calibre gun mounted in any operational German AFV.

Firing armour-piercing rounds at the typical combat range of 1000m (3280ft), the Jagdtiger could penetrate an impressive 230mm (9.1in) of vertical armour, enough to destroy any extant Allied AFV. The Jagdtiger's lethality, however, was undermined by its limited ammunition capacity of just 38 rounds.

Due to shortages in the 128mm (5in) gun, the last 26 Jagdtigers produced in early 1945 instead mounted the King Tiger's standard 8.8cm (3.4in) L/71 KwK 43 gun. In addition to its main armament, the Jagdtiger sported a standard ball-mounted 7.92mm (0.31in) MG 34 machine gun in the hull front for close-range anti-personnel defence purposes.

Jagpanther training
A Jagdpanther crosses a plain on a training exercise. This Jagpanther appears to be covered with *Zimmerit* anti-mine paste. *Zimmerit* was used to produce a hard layer covering the metal armour of the vehicle, providing enough separation that magnetically attached anti-tank mines would fail to stick.

Manned by a crew of six, the Jagdtiger had excellent survivability. The vehicle's frontal superstructure armour, sloped at 75 degrees, was an incredible 250mm (9.8in) thick, which rendered it invulnerable to frontal attack by any Allied tank or anti-tank gun. The tank destroyer's superstructure sides and rear, however, sported merely 80mm (3.14in)-thick steel plates. The gun's significant mass, its heavy rounds, and the thick frontal armour combined to make the Jagdtiger weigh a colossal 71.7 tonnes (70.6 tons) – the heaviest operational AFV of World War II. The vehicle was powered by the potent 522kW (700hp) Maybach HL230 engine utilized in the King Tiger tank. Given the vehicle's very large size and weight, even this potent powerplant enabled only a meagre maximum cross-country speed of 17km/h (11mph), although it could reach a reasonable top road speed of 38km/h (24mph). The vehicle's already limited mobility was further retarded by its enormous fuel consumption rates and the inability of most bridges to take its vast weight.

Jagdtiger

These views of the Jagdtiger heavy tank destroyer ably illustrate the monstrous size of its 128mm (5in) cannon as well as its tall, box-like, armoured superstructure.

Panzerjäger Tiger Ausf. B Jagdtiger
Crew: 6
Production: 1944–45
Weight: 71.7 tonnes (70.5 tons)
Dimensions: length: 10.65m (34ft 11in); width: 3.6m (11ft 10in); height: 2.8m (9ft 2in)
Engine: 522kW (700hp) Maybach HL230P30
Speed: 34km/h (21mph)
Range: 120km (75 miles)
Armament: 1 × 12.8cm (5in) PaK 44 L/55 main gun; 1 × 7.92mm (0.31in) MG 34
Armour: 80–250mm (3.14–9.84in)

The 77 Jagdtigers produced were allocated to three independent anti-tank units, including the 512th and 653rd Army Heavy Anti-tank Battalions. The nine operational Jagdtigers of the 653rd Battalion played a significant role in Operation 'Northwind', the January 1945 German offensive in Alsace-Lorraine. During these operations, one of the battalion's Jagdtigers was destroyed by Allied flanking fire near Rimling.

It is believed that not a single Jagdtiger was destroyed by enemy frontal fire in the entire war. However, Allied flanking hits and mobility kills, as well as abandonment due to running out of fuel, meant that by 29 April 1945 across all fronts just five Jadgtigers remained operational.

Jagdpanzer 38(t) Hetzer ('Baiter') light tank destroyer

In 1943, the Germans wrestled with the problem of designing a cheap and easy-to-build light tank destroyer to replace the increasingly non-survivable Marder series of improvised SPATGs then fielded by the anti-tank battalions of infantry divisions. The proven chassis of the obsolete Panzer 38(t) tank was still in production as a basis for several armoured support vehicles. This fact led to the development of the Hetzer ('Baiter') light tank destroyer on a widened version of this chassis.

Between March 1944 and April 1945, German armament firms completed 2584 Jagdpanzer 38(t) vehicles, making it the most common tank destroyer in

Jagtiger captured by US forces, 1945

The Germans abandoned many Jagdtigers due to lack of fuel or track damage; this Heavy Battalion 512 vehicle has spare track links fitted onto brackets on its turret sides and has most of its shallow armoured side skirts missing.

German service. Of advanced design technically, this small vehicle weighed just 16 tonnes (15.7 tons). It mounted a modified variant of the 7.5cm (2.95in) Pak 39 L/48 gun, the standard weapon installed in the Panzer IV tank. The Hetzer incorporated a steeply sloped, low silhouetted, armoured superstructure fitted directly onto

Jagdpanzer 38(t) Hetzer

This Hetzer is a mid- or late-production vehicle, identifiable by featuring just six lightening holes in its rear idler wheel rather than 12 holes, as well as sloping edges to the side-skirt front and rear corners.

Jagdpanzer 38(t) Hetzer

Crew: 4
Production: 1944–45
Weight: 16 tonnes (15.7 tons)
Dimensions: length: 6.2m (20ft 4in); width: 2.93m (9ft 7in); height: 1.96m (6ft 5in)
Engine: 112kW (150hp) Praga EPA/2 petrol
Speed: 35km/h (22mph)
Range: 214km (133 miles)
Armament: 1 x 7.5cm (2.95in) Pak 39 L/48 cannon; 2 x 7.92mm (0.31in) MG 34
Armour: 8–60mm (0.31–2.36in)

the widened Panzer 38(t) chassis. The 75mm (2.95in) gun was fitted in a limited-traverse Saukopf ('Pig's Head') mantlet positioned 38cm (15in) right of the centre line on the steeply sloped superstructure front plate. The small size of the Hetzer meant that the gun barrel considerably overhung the front of the vehicle. The vehicle also mounted a remotely controlled 360-degree-rotation 7.92mm (0.31in) MG 34 machine gun on the superstructure top operated by the crew commander.

By 1944 standards, the Hetzer possessed modest survivability with relatively thin plates; this was compensated for by the vehicle's low silhouette and well-sloped armour. The vehicle possessed 60mm (2.4in)-thick hull nose plates sloped at 40–60 degrees and 20mm (0.79in) superstructure side plates set at 20 degrees. In addition, the vehicle sported 5mm (0.19in)-thick side-skirts to ward off hollow charge round attacks against its tracks. The Hetzer was powered by the same 112kW (150hp) Czech Praga petrol engine fitted in the Panzer 38(t) tank, but the vehicle's increased weight (in comparison to the latter's) reduced its power-to-weight ratio to a mere 9.4hp/tonne. This restricted the Hetzer's maximum road speed to a relatively slow 26km/h (16mph) and just 15km/h (9mph) cross country.

Increasingly during 1944, the Hetzer replaced the remaining Marder vehicles in the self-propelled anti-tank companies of German infantry divisions. The 1944-format infantry division was authorized 10 Hetzers, while the new 1945-pattern infantry division was authorized 14 such vehicles in its retitled tank destroyer company. Due to production delays and transportation difficulties, only around 40 German divisions ever received their allotment of Hetzers; even then, they rarely received the full authorized strength. The vehicle also served in seven mechanized divisions, as well as five such brigades and seven independent units. This hard-to-locate and fuel-efficient small tank destroyer provided useful mobile anti-tank fire support primarily for German infantry during the war's final 18 months.

SELF-PROPELLED ARTILLERY VEHICLES

Within German mechanized formations there was a similar need for indirect fire support as that found in infantry divisions. The German Army started World War II with tractor-drawn artillery pieces providing the indirect fire support required by the highly mobile tank spearheads of panzer divisions. Often, however, tractor-drawn pieces could not keep within firing range of the vanguard tank squadrons as the latter advanced swiftly across rough cross-country terrain. Eventually the Germans recognized that what the armoured divisions needed was a fully tracked AFV that mounted an artillery piece as an integral part of the vehicle.

Despite the fact that such vehicles remained uncommon within the Wehrmacht, this section explores the three most important of such weapons: the Wespe ('Wasp'), Hummel ('Bumble Bee') and Grille ('Cricket') self-propelled artillery gun vehicles.

10.5cm (4.13in) leFH 18/2 auf Fahrgestell Panzer II (sf) Wespe ('Wasp')

Of the three principal German self-propelled artillery vehicles of World War II, the Wespe ('Wasp') was probably the best-known design. Its full designation, as above, literally meant: 105mm (4.13in) Light Field Howitzer 18/2 mounted on the self propelled Panzer II chassis 'Wasp'.

In 1942, three German armament firms – Alkett, MAN and Rheinmetall Borsig – worked together to design the vehicle; production, however, was carried out by the firm of FAMO, located in Warsaw in German-occupied Poland. The vehicle mounted the standard German light field piece, the ubiquitous 10.5cm (4.13in) le FH 18/2 light field howitzer – this variant being fitted with a muzzle brake. This gun delivered its high-explosive rounds with a terminal muzzle velocity of 470m/s (1542ft/s). The gun was mounted in a thinly armoured fighting compartment on top of a modified Panzer II chassis.

By 1940, the Panzer II was scarcely fit for combat, but its chassis was reliable and it made sense to utilize this for this new self-propelled gun design.

The Panzer II was a relatively small vehicle, which meant that the Wespe could carry only 40 105mm (4.13in) rounds, limiting its tactical

Wespe

The large single light located on the left side of the hull front is clearly evident in this Wespe vehicle; behind this, located in the side of the front superstructure, is a small horizontal vision slit for the driver.

10.5cm (4.13in) leFH18 auf PzKpfw II Wespe

Crew: 5

Production: 1943–44

Weight: 12.1 tonnes (11.9 tons)

Dimensions: length: 4.81m (15ft 10in); width: 2.28m (7ft 6in); height: 2.25m (7ft 4in)

Engine: 104kW (140hp) Maybach HL62TR

Speed: 40km/h (25mph)

Range: 220km (137 miles)

Armament: 1 x 105mm (4.13in) LeFH 18M L/28 howitzer

Armour: 5–30mm (0.19–1.18in)

Wespe fighting compartment
This side shot of a Wespe shows how tight space was inside the vehicle's fighting compartment. The space was open to the rear, giving the crew very little aerial protection.

utility. To compensate for this, every Wespe battery also contained a Wespe munitions carrier that stowed an additional 90 rounds of ammunition. German firms constructed 158 of these carriers by utilizing either turret-less Panzer II tanks or gun-less Wespe vehicles.

The howitzer was mounted within an open-topped, box-like superstructure that sloped downwards towards the rear; there was thus merely a low rear plate to this superstructure. This superstructure was built directly upon the Panzer II chassis. The vehicle, served by a crew of five, had modest protection: 18mm (0.7in)-thick armour on the hull and 10mm (0.39in)-thick superstructure plates.

Both the modest survivability provided and the few rounds carried enabled the vehicle to weigh no more than 11.5 tonnes (11.3 tons), so as not to overload the Panzer II chassis.

Powered by a 104kW (140hp) Maybach engine, the vehicle featured a modified Panzer II suspension; these enabled the Wespe to obtain a maximum by-road speed of 40km/h (25mph) and a disappointing 20km/h (12mph) cross country.

Between 1942 and late 1944, FAMO produced 684 Wespe vehicles, either by converting obsolete Panzer II tanks or by utilizing newly fabricated Panzer II chassis. The Wespe self-propelled gun served in the armoured artillery battalions found within German panzer and panzergrenadier (mechanized) divisions. These battalions typically fielded two Wespe-equipped light batteries, each with six vehicles. Given the limited number produced, the significant number of such battalions, and combat casualties, many Wespe batteries remained short of their authorized strength of six vehicles.

15cm (5.9in) PzFH 18/1 auf Geschützwagen III/IV (sf) Hummel ('Bumble-Bee')

Designed by the firm of Alkett in 1942, the Hummel ('Bumble Bee') was a heavy self propelled artillery vehicle that equipped the solitary heavy battery in the panzer division armoured artillery battalion. The full official German designation translates as 'Armoured Howitzer 18/1 on the self-propelled Gun Carriage III/IV'.

The vehicle sported the 15cm (5.9in) sFH 18/1 L/30 heavy field howitzer – the standard German heavy field piece. The Hummel, which was serviced by a six-man crew, comprised this large gun mounted in a lightly armoured superstructure that was fitted on top of the hybrid Geschützwagen III/IV chassis, which incorporated features of both the Panzer III and Panzer IV tank designs. In reality, the chassis was that of a Panzer IV with a front-positioned Maybach engine, which also incorporated the final drive and sprockets of the Panzer III tank. Some of the very last Hummel vehicles to be manufactured, however, utilized the standard Panzer IV chassis due to a shortage of these hybrid chassis. The vehicle also possessed a 7.92mm (0.31in) MG 34 machine gun for close-protection purposes, for which it carried 600 rounds.

This large gun ensured that the vehicle weighed a hefty 25.9 tonnes (25.5 tons). When designing the Hummel, Alkett had to be careful not to overload the bearing capability of the Hybrid III/IV chassis; the main way in which the vehicle's weight was kept down was to stow on board only a modest 18 150mm (5.9in) rounds. This restricted the vehicle's tactical utility, although this was in part compensated for by the provision of one Hummel munitions carrier for each battery. The firm of Deutsche Eisenwerke constructed 150 of these

Hummels, Russia, summer 1943
A battery of heavily-camouflaged Hummel vehicles sit deployed in open scrub terrain; the crews' apparently relaxed countenance seems to suggest that no tactical engagement is imminent.

15cm (5.9in) schwere Panzerhaubitze Hummel
Crew: 6
Production: 1943–45
Weight: 25.9 tonnes (25.5 tons)
Dimensions: length: 7.17m (23ft 6in); width: 2.97m (9ft 8in); height: 2.81m (9ft 2in)
Engine: 224kW (300hp) Maybach HL120TRM
Speed: 42km/h (26mph)
Range: 215km (134 miles)
Armament: 1 x 150mm (5.9in) sFH 18/1 L/30 howitzer; 1 x 7.92mm (0.31in) MG
Armour: 10–20mm (0.39–0.79in)

Hummel
The slightly sloping armoured vision slit in this Hummel's front left superstructure is evident; this and a similar one on the right side provided additional situational awareness for the vehicle's driver and radio operator.

carriers, which were simply gun-less Hummel vehicles. The second way the vehicle's weight was kept down was by providing modest armour protection: a mere 20mm (0.79in)-thick plate on the lower hull and just 10mm (0.39in) superstructure armour. The Hummel could obtain a maximum by-road speed of 42km/h (26mph) and an operational range of 215km (134 miles) when travelling by road.

Construction of the Hummel was undertaken entirely by Deutsche Eisenwerke. During a 19-month production run that ran from December 1942 through to June 1944, the firm constructed 669 Hummels at an average rate of 35 units per month. Six Hummel vehicles equipped the solitary heavy self propelled battery within each armoured artillery battalion that the panzer divisions possessed. The Hummel first saw combat in significant numbers in Operation 'Citadel', the July 1943 German armoured onslaught against the Soviet-held Kursk salient in the centre of the Eastern Front. Around 100 Hummel vehicles, deployed among 22 German mechanized formations, participated in this offensive, where their heavy firepower was urgently needed to help penetrate the well-established Soviet defensive belts. The Hummel served in smaller numbers across all fighting fronts right up until the war's final days, by which time only a few dozen remained operational.

15cm (5.9in) SiG 33 (sf) auf Panzer 38(t) Ausf H Grille ('Cricket')

Designed during 1942, the Grille ('Cricket') was a heavy self-propelled gun vehicle that mounted the 15cm (5.9in) SiG 33 heavy infantry gun on the chassis of the by-then obsolete Czech-designed Panzer 38(t) tank. The Model H Grille vehicle had a shallow open-topped armoured fighting compartment built on top of the rear-engined Panzer 38(t) Model H chassis. The Grille enjoyed a reasonable degree of survivability, with armoured plates up to 50mm (1.96in) thick on the hull front and 25mm (1in)-thick superstructure plates.

The firm of BMM in Prague manufactured 200 Grille vehicles during a six-month production run from February through to June 1943; the firm also produced a final batch of 10 Grille vehicles in November 1943. The vehicle served in the infantry gun companies of German panzergrenadier regiments found within armoured and mechanized divisions. Each company had an authorized strength of six Grille vehicles, although this was not commonly achieved due to the small numbers produced.

A second Grille variant emerged in 1943: the Model K. This mounted the same 15cm (5.9in) SiG 33 heavy infantry gun on the fundamentally redesigned Panzer 38(t) Model M chassis; the latter had been purpose-built as a platform for various self-propelled weapons, including the Hetzer light tank destroyer. This chassis had the 112kW (150hp) Praga engine moved to the centre rather than the rear; this allowed the open-topped fighting compartment to be installed on top of the rear of the chassis. The open-topped superstructure on the Model K was smaller but higher than its predecessor, the Model H Grille. German firms manufactured 162 Model K vehicles during a 10-month production run that started in December 1943.

Sdkfz 138 'Grille' Ausf H

The Model H Grille vehicle, seen here, looked very different from its sister vehicle (the Model K, opposite, top); its 15cm (5.9in) sIG 33 infantry gun was forward-located within a simple box-like, sloping superstructure.

Sdkfz 138 'Grille' Ausf K

The hinged plate attached to the bottom of the gun of this Model K Grille moved up as the gun was elevated to fill the ensuing gap in the fighting compartment's front plate.

15cm (5.9in) schwere Infanterie Geschütz auf Selbsfahrlafette 38(t) Ausf K Grille
Crew: 5
Production: 1943–44
Weight: 12.7 tonnes (12.4 tons)
Dimensions: length: 4.61m (15ft 1in); width: 2.16m (7ft); height: 2.40m (7ft 10in)
Engine: 112kW (150hp) Praga EPA/2
Speed: 35km/h (22mph)
Range: 185km (115 miles)
Armament: 1 x 15cm (5.9in) sIG 33 gun; 1× 7.92mm (0.31in) MG 34
Armour: 50mm (1.96in) hull front

Both Grille variants suffered from the same problem as the Wespe and Hummel: limited main gun stowage – the Model K Grille vehicle carried a mere 15 rounds for its main armament. The Germans adopted the same compensatory method as developed for the other two self-propelled gun vehicles. Between January and May 1944, the firm of BMM constructed 102 gun-less Model K Grille vehicles to serve as munitions carriers. These vehicles were also provided with a field conversion kit that meant they could be swiftly converted into the standard gunned Grille vehicle at the front line if needs be, for example, to replace destroyed gunned vehicles.

German self-propelled gun lethality compared

Vehicle	Wespe	Hummel	Grille
Main gun muzzle velocity	470m/s (1542ft/s)	520m/s (1706ft/s)	240m/s (787ft/s)
Maximum range	10,675m (35,020ft)	13,250m (43,470ft)	4700m (15,420ft)

ARMOURED CARS AND HALF-TRACKS

In addition to tanks and self-propelled guns, another genre of AFVs made an important contribution to the combat effectiveness generated by Germany's wartime divisions. This genre included wheeled armoured cars and half-tracked armoured personnel carriers (APCs). Developed in the inter-war period, German wheeled armoured cars performed reconnaissance, intelligence-gathering and scouting duties. German half-tracked APCs, meanwhile, were designed to transport infantry personnel to the edge of the tactical battlefield, where they would debus and join combat. As World War II unfolded, however, the tactical employment of these AFVs mutated. Increasingly, these vehicles became embroiled in the tactical battle or else conducted reconnaissance by force. This led to the development of numerous up-gunned and other specialized APC variants and a new generation of heavily gunned wheeled armoured cars.

This secion examines the eight most important examples of this AFV genre.

Sdkfz 231 six-wheeled heavy armoured car

The first series of armoured cars designed by the inter-war German Army was the six wheeled Sdkfz 231 heavy reconnaissance vehicles. These four-man vehicles were based around a Büssing-NAG chassis with two sets of rear-drive wheels and a single pair of steering front wheels. On top of

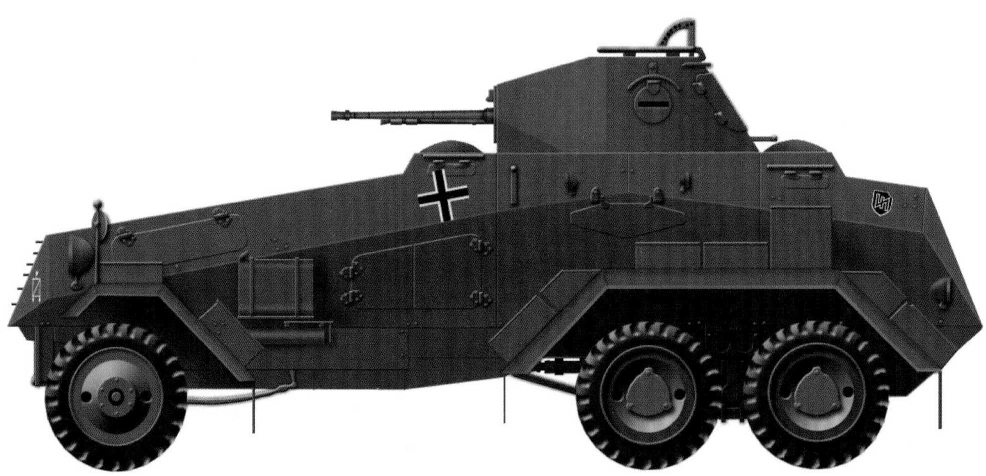

Sdkfz 231 6-Rad armoured car

This left-side view of a Sdkfz 231 six-wheeler armoured car reveals the left-mounted turret machine gun and the centre-mounted 20mm cannon; the armoured vision slit in the turret sides is also visible.

SdKfz 231 Schwerer Panzerspähwagen 6-Rad

Crew: 4
Production: 1930–36
Weight: 7.9 tonnes (7.8 tons)
Dimensions: length: 5.57m (18ft 7in); width: 1.82m (5ft 11.5in); height: 2.25m (7ft 4.5in)
Engine: 115kW (155hp) Daimler-Benz, Büssing-NAG
Speed: 70km/h (43mph)
Range: 300km (186 miles)
Armament: 1 x 20mm (0.79in) KwK 30 cannon; 1 x 7.62mm (0.3in) co-axial MG
Armour: 8–15mm (0.31–0.59in)

the chassis was constructed a lightly armoured superstructure. Positioned towards the back of the vehicle, on the superstructure roof, sat a small angular turret. This mounted a 2cm (0.79in) KwK 30 cannon as well as a co-axial 7.92mm (0.31in) Mauser MG 13 machine gun; later vehicles instead mounted a co-axial MG 34. The vehicle weighed 7.9 tonnes (7.8 tons) and was powered by a 70hp Magirus engine.

Between 1932 and 1935, Daimler-Benz manufactured 123 Sdkfz 231 six-wheeled half-tracks. Establishing a practice that would be followed by subsequent series of armoured cars, the Germans built specialized variants of the six-wheeler Sdkfz 231, which confusingly were given different Sdkfz designations. The Sdkfz 232 radio car, for example, was a command variant that incorporated a powerful 100-Watt radio transmitter serviced by a distinctive horizontal frame antenna.

Sdkfz 221 four-wheeled light armoured car

The second series of armoured cars developed by the inter-war German Army was the Horch-designed Sdkfz 221 four-wheeled light reconnaissance vehicles. Entering production in 1935, this 3.75 tonne (3.7 ton) compact two-man armoured car mounted a single 7.92mm (0.31in) MG 34 machine gun and was protected by 8–14.5mm (0.31–0.57in)-thick armoured plates. Thanks to its 56kW (75hp) Horch petrol engine, the Sdkfz 221 could

SdKfz 221 Leichte Panzerspähwagen MG
Crew: 3
Production: 1935–44
Weight: 3.75 tonnes (3.7 tons)
Dimensions: length: 4.80m (15ft 8in); width: 1.95m (6ft 5in); height: 1.70m (5ft 7in)
Engine: 56kW (75hp) Horch 3.5 L petrol
Speed: 80km/h (50mph)
Range: 198km (124 miles)
Armament: 1 x 7.92mm (0.31in) MG
Armour: 8–14.5mm (0.31–0.57in)

Sdkfz 221 armoured car
This left-hand image of a standard MG 34-equipped Sdkfz 221 scout car reveals just how shallow the vehicle's turret was; note the large double-hinged hatch on the left rear superstructure.

obtain an impressive road speed of up to 80km/h (50mph). Fitted with a 100-litre (22-gallon) fuel tank the armoured car enjoyed an impressive operational radius of 198km (124 miles) across rough terrain.

As per the usual practice, the Germans developed several variants of this vehicle, including a radio car version. The four-wheeled Sdkfz 222, moreover, was a heavier 4.8 tonne (4.7 ton) version of the Sdkfz 221 that carried the 2cm (0.79in) KwK 30 cannon and co-axial 7.92mm (0.31in) MG 34 mounted in a distinctive 10-sided, open topped, fully-rotating turret.

In total during 1935–42, the Germans manufactured 2116 of these Horch-designed light armoured cars.

Sdkfz 231 eight-wheeled heavy armoured car

The third and final series of armoured cars developed by the inter-war German Army was the confusingly designated Büssing NAG Sdkfz 231 eight wheel heavy vehicles, of which 1235 were manufactured in 1937–42. Visually, these vehicles looked similar to the Sdkfz 231 six wheelers, except that these longer eight-wheeler armoured cars mounted the identical turret further forwards, over the second wheeled axle. Causing much confusion, German firms also constructed eight-wheeler Sdkfz 232 and 263 signals cars and Sdkfz 233 heavy support variants of these Büssing-NAG Sdkfz 231 eight-wheeler vehicles.

The standard four-man Sdkfz 231 eight-wheeler car weighed 8.4 tonnes (8.3 tons) and mounted the 2cm (0.79in) KwK 38 gun together with an MG 34 machine gun; the earliest vehicles, however, sported the KwK 30 cannon and the Mauser MG 13 machine gun.

These eight-wheeled armoured cars possessed driving controls at both ends of the vehicle. The vehicles sported eight separately suspended and separately steered wheels, meaning several could be damaged yet the vehicle would retain its mobility; this innovation made these vehicles some of the most advanced armoured cars of the pre-war era.

Sdkfz 231 8-Rad, winter camouflage

Painted in white winter camouflage, this armoured car fought at Kharkov with 3rd SS Panzergrenadier Division *Totenkopf* during the late winter and early spring of 1943.

Sdkfz 231 8-Rad armoured car
As the extensive frame aerials make obvious, this is the signal car ('FuG') variant of the Sdkfz 231 eight-wheeled armoured car (which the Germans confusingly designated the Sdkfz 232).

SdKfz 231 Schwere Spähpanzerwagen 8-Rad
Crew: 4
Production: 1937–42
Weight: 8.4 tonnes (8.3 tons)
Dimensions: length: 4.67m (15ft 4in); width: 2.2m (7ft 2in); height: 2.35m (7ft 8in)
Engine: 157kW (210hp) Büssing-NAG L8V
Speed: 85km/h (53mph)
Range: 300km (186 miles)
Armament: 1 x 2cm (0.79in) KwK 30/38 L/55 cannon; 1 x 7.92mm (0.31in) co-axial MG 13
Armour: 8–15mm (0.31–0.59in)

Sdkfz 234/1 eight-wheeled heavy armoured car

In 1942, the German Army recognized that only a new generation of better-protected and more powerfully armed armoured cars could perform reconnaissance effectively on the current tactical battlefield. This led to the development during 1943 of the Sdkfz 234 eight-wheeled heavy armoured cars, of which four variants were constructed. The key feature of these new armoured cars was that each wheel had independent drive. The

Germans had learned through costly previous losses that the earlier wheeled armoured cars might become disabled if even one wheel was damaged. With the Sdkfz 234 armoured cars, the vehicle could maintain normal tactical mobility even after several wheels had been damaged.

The basic Sdkfz 234/1 weighed 11.7 tonnes (11.6 tons) – heavier than any previous German armoured car. Despite its bulk, the vehicle's powerful 156kW (210hp) diesel engine enabled it to obtain an impressive top by-road speed of 88km/h (55mph). The Sdkfz 234/1 armoured car mounted a 20mm (0.79in) cannon as its main armament. In 1943–44, German firms manufactured 708 Sdkfz 234/1s, which were issued to panzer division armoured car companies. In combat, however, the vehicle's 20mm (0.79in) gun proved inadequate to enable it to fight in order to gather intelligence, which was the method increasingly favoured by German tactical doctrine.

Sdkfz 234/2-4 Puma eight-wheeled heavy armoured cars

By 1943, the Germans desperately needed greater firepower at the front. This led to the development of the Sdkfz 234/2 'Puma' eight-wheeled heavy armoured car. The vehicle mounted the long barrelled 5cm (1.96in) KwK 39/1 L/60 gun in a fully revolving turret, rather like a wheeled light tank. The combined weight of the larger-calibre gun and the turret resulted in a modest decline in the top speed and maximum operational range that the Puma could obtain. However, these limitations were more than compensated for by the vehicle's proven anti-tank ability against lightly armoured enemy AFVs.

From late 1943, the Puma began to join the armoured reconnaissance battalion found within panzer and panzergrenadier divisions. Initially during 1943 it was hoped that the Puma might completely replace the Sdkfz 234/1. The Puma remained a rare vehicle, however, with just 101 vehicles being completed. As a result, both the Sdkfz 234/1 and the Puma continued to serve alongside each other until the end of the war.

The combat effectiveness that the Puma demonstrated led the Germans to develop the improvised Sdkfz 234/3. This 9.95-tonne (9.8-ton) variant of the Puma mounted the short-barrelled 7.5cm (2.95in) KwK

Sdkfz 234 'Puma' variants compared

Variant	Production	Vehicles produced	Arrangement for mounting main gun
234/2	1943–44	101	Fully rotating turret
234/3	1944–45	88	Within open-topped superstructure
234/4	1944–45	89	Within open-topped superstructure

L/24 low-velocity gun formerly mounted in early-war Panzer IV tanks. This limited-traverse gun sat within an open-topped superstructure, which afforded the crew only modest protection. The final vehicle in the Puma series, the Sdkfz 234/4, entered German service in 1944. It sported the longer-barrelled and potent 7.5cm (2.95in) KwK L/48 of the Panzer IV Models G–J tank, once again mounted directly within an open-topped superstructure. Between June 1944 and March 1945, German factories produced 88 Sdkfz 234/3 and 89 Sdkfz 234/4 armoured cars.

SdKfz 234/2 Puma
Crew: 4
Production: 1943–45
Weight: 10.5 tonnes (10.3 tons)
Dimensions: length: 6.02m (19ft 9in); width: 2.36m (7ft 9in); height: 2.1m (6ft 10in)
Engine: 157kW (210hp) Tatra 103 V-12 diesel
Speed: 90km/h (55mph)
Range: 1000km (625 miles)
Armament: 1 x 5cm (1.96in) KwK 39/1 L/60 gun cannon; 1 x 7.92mm (0.31in) MG 34 co-axial
Armour: 9–30mm (0.35–1.18in)

SdKfz 234/2 Puma armoured car
This Puma, bedecked in three-tone camouflage, sports a spare wheel carried on its hull rear plate; note the three smoke grenade launchers fitted to the forward part of the vehicle's turret side.

SS half-tracks and StuGs at Kursk, July 1943
An MG42-equipped Sdkfz 251 APC transports troops from the 2nd SS Panzergrenadier Division *Das Reich* across open terrain during the July 1943 'Citadel' offensive at Kursk.

Sdkfz 251 medium half-tracked armoured personnel carrier

During the late 1930s, the Germans developed vehicles that eventually took on many of the functions intended for the armoured car. These vehicles were half-tracked armoured personnel carriers (APCs), or Schützenpanzerwagen (SPW). The APC became the vehicle of choice for the protected transport of infantry up to, and within, the tactical battlefield. The creation of Germany's first panzer divisions in 1935 created the requirement for an armoured vehicle that could transport the infantry (panzergrenadiers) that would fight alongside the tanks.

Companies constructing the Sdkfz 251	
Firm	**Sdkfz 251 models constructed**
Deutsche Werke	D
Hanomag (Hannover)	A–C
MNH (Hannover)	D
Schichau (Elbing)	A–C
Weserhuette (Bad Oeynhausen)	A–C
Wumag (Goerlitz)	A–C

In 1935, the German Army decided that the chassis of the existing 3 tonne (2.95 ton) half-track tractor was, with modest modification, suitable to transport the standard 10-man infantry squad. The ensuing vehicle, the Sdkfz 251 mittlerer (medium) SPW (mSPW), entered production in mid-1939. By the time that World War II erupted in September, 67 Sdkfz 251s were in service.

To this tractor chassis the Germans added an open-topped, well sloped armoured superstructure that had rear doors; the infanteers could debus from the vehicle either through these doors or by jumping out over the vehicle's superstructure sides. The sloped superstructure featured 14.5mm (0.57in) frontal armoured plates and 8mm (0.31in)-thick side/ rear plates. The chassis sported a single front-located wheeled axle plus a rear-positioned tracked arrangement of seven interleaved road wheels and a separate front wheel. The vehicle mounted an MG 34 without a shield at the front of the open-topped fighting compartment. This original Model A design was distinguishable from its successor variants by its three prominent rectangular vision ports positioned along the top of each side of the superstructure. The small number of mSPWs in service in the 1939 Polish campaign served in German panzer divisions alongside the more numerous lightly armoured lorries that transported the divisions' infantry element. As originally designed, the role of the mSPW was to transport the infantry to the edge of the tactical battlefield, where they would debus and participate in combat on foot. The Sdkfz 251, therefore, was optimized for fast cross-country performance (to keep up with the panzers) at the price of being only lightly armoured.

Some 347 Sdkfz 251 Model B and C vehicles were produced in 1940. The design of the Model B deleted the side vision ports, in order to simplify production, but added a distinctive small splinter shield fitted to the forward-located MG 34 machine gun to provide the gunner with additional protection. The Model C also incorporated modifications designed to simplify and speed up construction, including a

SdKfz 251/1 Mittlere Schützenpanzerwagen Ausf B
Crew: 2, plus 10 troops
Production: 1939–45
Weight: 9.9 tonnes (9 tons)
Dimensions: length: 5.98m (19ft 7in); width: 2.1m (6ft 11in); height: 1.75m (5ft 8in)
Engine: 74kW (99hp) Maybach HL42TUKRM
Speed: 53km/h (33mph)
Range: 300km (186 miles)
Armament: 2 x 7.62mm (0.3in) MG 34
Armour: 8–14mm (0.31–0.57in)

Sdkfz 251/1 Ausf B half-track
This standard troop-transporting APC, the early Sdkfz 251/1 Model B, features two MG 34s (with the forward weapon fitted with a splinter shield). The Model A was identical save for three vision slits in the superstructure sides.

single-piece hull nose front plate rather than the two-piece angled nose sported by the Model A and B vehicles. Large-scale production of the further simplified Model D ensued during 1941–44. Peak manufacture occurred in 1944, when 7820 mSPWs were constructed. By the war's end, some 16,300 mSPW of all types had been produced.

During 1941–42, the Sdkfz 251 started equipping a single battalion within one of the German panzer division's subordinate panzergrenadier regiments. This SPW-equipped unit became known as the *gepanzerte* battalion and had an authorized strength of 160 Sdkfz vehicles. There were simply an insufficient number of mSPWs to equip more than one battalion in the majority of German panzer and panzergrenadier divisions. The total number of Sdkfz 251 half-track APCs in German service peaked at 6147 in December 1944.

Sdkfz 251 medium half-tracked APC variants

During both the 1940 Western campaign and the summer 1941 Axis invasion of the Soviet Union, German tactical employment of APCs developed rapidly. Increasingly, APCs were moving into the heart of intense combat actions, where their embussed infanteers would often join the firefight while still located within their vehicles. These tactical changes led the Germans to develop 22 variants of the basic Sdkfz 251, some being better-armed and armoured versions of the original.

The first variant, the Sdkfz 251/1, sported a quadruple 7.92mm (0.31in) MG 34 machine gun mounting. Some Sdkfz 251/1 vehicles were modified to fire six 28cm/32cm (11–12.6in) Wurfkorper rockets from frame mountings fitted to the vehicle's sides. Entering production in 1941, the Sdkfz 251/10 was the first up-gunned

Panzergrenadiers, winter, Russian Front

In the background two Sdkfz 251 half-tracks carry a section of German troops through severe wintry conditions in the Soviet Union, their tracks throwing up ice and snow as they advance; note the seven interleaved road wheels and driver wheel of the rear tracked arrangement.

Sdkfz 251/9 haltrack
The left-hand rear door set in the armoured superstructure of this 75mm (2.95in)-gunned Sdkfz 251/9 hangs open, illustrating how troops entered or exited the vehicle.

SdKfz 251/9 Mittlere Schützenpanzerwagen 7.5cm (2.95in) Ausf D
Crew: 3
Production: 1943–45
Weight: 9.4 tonnes (9.25 tons)
Dimensions: length: 5.98m (19ft 7in); width: 2.83m (9ft 4in); height: 2.07m (6ft 10in)
Engine: 74kW (100hp) Maybach HL42TUKRM
Speed: 53km/h (33mph)
Range: 300km (186 miles)
Armament: 1 x 75mm (2.95in) KwK 37 L/24 gun
Armour: 5.5–14.5mm (0.21–0.57in)

variant; it mounted a 3.7cm (1.46in) Pak 36 anti-tank gun on the forward superstructure.

The Sdkfz 251/9, introduced in 1942, similarly sported the short-barrelled 7.5cm (2.95in) KwK 37 L/24 calibre gun formerly mounted in early Panzer IV tanks. The final up-gunned variant, the Sdkfz 251/22, even carried the complete 7.5cm (2.95in) Pak 40 anti-tank gun. Another offensively orientated firepower variant was the Sdkfz 251/16, which carried two 700 litre (154-gallon) flame fuel tanks and two 14mm (0.55in) flame projectors mounted on the superstructure top on either side of the vehicle.

Other mSPW variants had a more defensive orientation. Two models introduced during 1943–44, for example, were intended to help counter the growing Allied tactical air threat. The Sdkfz 251/17 sported a 2cm (0.79in) Flak 30 or 38 anti-aircraft gun on its superstructure front. The Sdkfz 251/21, moreover, incorporated a treble mounting for the Luftwaffe 15mm (0.6in) MG 151 aircraft machine gun. Finally, there were a series of mSPW variants designed for specific tactical roles. Such vehicles included the Sdkfz 251/5 armoured engineer vehicle and the Sdkfz 251/6 senior leader's command APC.

Sdkfz 250 light half-tracked APC

In 1939, the Germans developed a small light half-tracked APC vehicle using the chassis of the extant Sdkfz 10 artillery tractor. This new troop transporter was designated the Sdkfz 250/1 leichter (light) SPW (leSPW).

The Sdkfz 250/1 was smaller than the 251 and featured a scaled down version of the latter's open-topped armoured superstructure. The vehicle had a tracked rear section with five interleaved road wheels and a separate front wheel, together with the front-wheeled single axle. Crucially, the Sdkfz 250 could carry only six personnel; the vehicle was too small to move a complete infantry section, the basic micro-tactical unit. Despite its limited size, however, the standard leSPW was well suited for a variety of specialized roles such as an engineering, a reconnaissance or a mortar vehicle.

The basic Sdkfz 250/1 vehicle weighed 5.3 tonnes (5.2 tons) and could reach an impressive top speed of 60km/h (37mph). The leSPW sported 14.5mm (0.57in)-thick frontal armour as well as 8–10mm (0.31–0.39in)-thick plates on the rear and sides. The Sdkfz 250 usually carried one or two 7.92mm (0.31in) MG 34 machine guns that were fitted onto pivot mounts located at the front and sometimes the rear of the superstructure. Production of the leSPW commenced at the Demag factory in December 1939. The first completed vehicles entered front-line German service in February 1940. These first vehicles were allocated to the commanders of panzergrenadier companies or platoons as their Headquarters vehicle.

The Sdkfz 250 first saw combat in the May 1940 Western campaign,

Sdkfz 250/3 half-track

The Sdkfz 250/3 signals vehicle featured an extensive frame antenna; its four sub-variants (3-I to 3-IV) mounted different radio sets depending on its roles as an air operations, close air support or ground assault vehicle.

SdKfz 250/3 Leichte Schützenpanzerwagen

Crew: 2, plus 4 troops
Production: 1941–45
Weight: 5.3 tonnes (5.2 tons)
Dimensions: length: 4.56m (14ft 11in); width: 1.95m (6ft 5in); height: 1.66m (5ft 5in)
Engine: 74kW (100hp) Maybach HL42TRKM 6-cylinder
Speed: 60km/h (37mph)
Range: 300km (186 miles)
Armament: 2 x 7.92mm (0.31in) MG 34
Armour: 5.5–14.5mm (0.21–0.57in)

Sdkfz 250/9 half-track

The Sdkfz 250/9 light reconnaissance/scout half-tracked vehicle, which was fitted with a Fu8 radio set, entered service during 1943. This is finished in an unusual olive and khaki bi-tone camouflage scheme.

SdKfz 250/9 Leichte Schützenpanzerwagen

Crew: 3
Production: 1941–45
Weight: 6.9 tonnes (6.3 tons)
Dimensions: length: 4.56m (14ft 11in); width: 1.95m (6ft 5in); height: 2.16m (7ft 1in)
Engine: 74kW (100hp) Maybach HL42TRKM 6-cylinder
Speed: 60km/h (37mph)
Range: 320km (199 miles)
Armament: 1 x 2cm (0.79in) KwK 38 autocannon; 1 x 1 x 7.92mm (0.31in) co-axial MG
Armour: 5.5–14.5mm (0.21–0.57in)

where it performed its allocated tactical roles in an adequate fashion. As greater demand for this light APC emerged from the field, production of the vehicle increased in 1941–42, with Adlerwerke, Büssing-NAG and MWC joining the manufacturing process. From 1943 onwards, this led to the design being regularly modified so as to simplify and thus speed up production of the vehicle. These simplifications included the replacement of armoured vision flaps with open slits. Production continued until the war's end, with 7232 Sdkfz 250s of all 15 variants being produced. Always less common that its larger cousin, in November 1944, for example, only 2185 Sdkfz 250s were in service across the entire German Armed Forces.

Sdkfz 250 light half-tracked APC variants

As the war unfolded, as we have seen, German tactical doctrine for the employment of APCs gradually developed. Just as with its larger cousin, the Sdkfz 250/1 APC found itself increasingly in the thick of fierce tactical combat. This led the Germans to develop a range of variants that either were up-gunned models or else fulfilled a variety of specialized roles. In addition to the basic vehicle, German armament firms constructed some 14 variants of the Sdkfz 250; some models duplicated the niche tactical roles performed by the 22 variants of the larger Sdkfz 251 medium APC.

Several variants brought much-needed firepower to the tactical

Sdkfz 250/10 half-track

This left-side view of a Sdkfz 250/10 shows how the 3.7cm Pak 35/36 was mounted, with three-sided splinter shield, on the vehicle's front superstructure. This vehicle has been finished in white-wash winter camouflage.

SdKfz 250/10 Leichte Schützenpanzerwagen
Crew: 4
Production: 1941–45
Weight: 6.3 tonnes (6.2 tons)
Dimensions: length: 4.56m (14ft 11in); width: 1.95m (6ft 5in); height: 1.97m (6ft 6in)
Engine: 74kW (100hp) Maybach HL42TRKM
Speed: 60km/h (37mph)
Range: 320km (199 miles)
Armament: 1x 3.7cm (1.46in) Pak 35/36 anti-tank gun
Armour: 5.5–14.5mm (0.21–0.57in)

battlefield. The Sdkfz 250/8, which entered service in 1943, for example, mounted a short-barrelled 7.5cm (2.95in) KwK 37 L/24 gun with a 7.92mm (0.31in) MG 34 located above it; the MG 34 could fire tracer and thus could act as a sighting/ranging device for the main gun. The Sdkfz 250/7 was a dedicated mortar carrier. The mortar firing team could fire the 81mm (3.2in) weapon from inside the vehicle, although often they preferred to deploy the mortar to a well-covered location and fire it from there.

Six Sdkfz 250/7 vehicles equipped the heavy gun platoon of a small number of elite panzergrenadier divisions. The Sdkfz 250/9, meanwhile, mounted the entire turret of the Sdkfz 222 armoured car on top of its roofed-in superstructure; this turret mounted a 20mm (0.79in) cannon. This variant was intended to replace the Sdkfz 222 armoured car, as part of a wider process whereby the Germans gradually replaced the more vulnerable and less mobile wheeled armoured cars with half-tracked Sdkfz 250 and 251 vehicles.

Several other leSPW vehicles were designed to augment front-line German anti-tank capabilities. The Sdkfz 250/10, for example, mounted the 3.7cm (1.46in) Pak 36 anti-tank gun. A similarly roled vehicle was the Sdkfz 250/11, which was introduced into German service in 1942. This vehicle carried the 2.8cm (1.1in) schwere Panzerbüsche 41 taper bore heavy anti-tank rifle mounted on the superstructure front. This anti-tank rifle delivered its PzGr Patrone 42 tungsten-core round at an incredible muzzle velocity of 1402m/s (4600ft/s). At a combat range of 500m (1640ft), this weapon could penetrate 66mm (2.59in) of vertical armoured plate.

Many other Sdkfz 250 variants were designed for specialized niche roles on the tactical battlefield. The Sdkfz 250/3 leichter Funkpanzerwagen, for example, was a radio-equipped command vehicle. It could readily be identified from its sister light half-tracks by the distinctive frame aerial along its superstructure. The vehicle was often employed by senior commanders; General Erwin Rommel used one as his personal command vehicle when he was Commander of the Afrika Korps during the 1942 North African campaign.

Eventually, late-production Sdkfz 250/3 models were fitted with vertical pole antennae with a small star aerial fitted at the top, rather than the bulky horizontal frame antenna. The remaining six Sdkfz variants not yet discussed were employed either as observation, ammunition carrying, or field telephone laying vehicles.

Production of the Sdkfz 250	
Year	**Vehicles produced**
1940–41	1030
1942	1337
1943	2895
1944	1701
1945	269

ARMOURED TRACTORS AND TRUCKS

There was a final genre of armoured support vehicles that played an important part in helping to create the combat effectiveness generated by German front-line formations during World War II. This was the category of 'prime-movers': wheeled, half-tracked and fully tracked vehicles used for towing and moving personnel, weapons, equipment and stores. This genre included half- and fully tracked armoured tractors and wheeled trucks. This chapter discusses the six most important examples of this category of hard-working yet often forgotten armoured support vehicles, including two half-tracked armoured tractors, one fully-tracked armoured tractor and three-wheeled trucks.

Sdkfz 7 medium tractor

The firm of Kraus-Maffei in Munich designed the Sdkfz 7 medium tractor in 1934, and limited construction commenced of several pre-production series. Larger-scale production of the final production variant, however, began only in 1939. The standard Sdkfz 7 weighed 9.2 tonnes (9 tons) and was driven by a 104kW (140hp) Maybach 6-cylinder engine that provided a top road speed of 50km/h (31mph). The vehicle could tow an 8-tonne (7.9-ton) payload – enough to transport the 15cm (5.9in) sFH 18 medium howitzer or the various 88mm (3.4in) Flak gun designs.

The Germans developed several variants of the tractor during the war. The Sdkfz 7/1 mounted the 2cm (0.79in) Flakvierling 38 quadruple anti-aircraft gun in the rear compartment.

Most of these vehicles featured a fully enclosed and armoured driver's cab. This vehicle served in the anti aircraft companies of panzergrenadier regiments to provide protection against low flying Allied tactical air attack. The similar Sdkfz 7/2 instead mounted the 3.7cm (1.46in) Flak 36, 37 or 43 gun directly on the floor of the rear compartment. The Germans produced 12,187 Sdkfz 7 tractors of all types in 1939–45.

Opel Blitz 3.6-36 medium truck

The Opel Blitz ('Lightning') 3.6-36 medium truck was the prime mover most widely employed by the German Armed Forces during the 1939–45 war. The chassis consisted of a single-wheeled axle at the front and the rear with a wide sweeping mudguard over the former, and with paired rear tyres. On top of this was a forward-positioned engine and windowed metal driving cab together with a large wooden flatbed rear section with shallow wooden sides. Many trucks had a canvas awning over the rear section.

The standard Blitz truck weighed 2.7 tonnes (2.6 tons) and was powered by an Opel 48kW (65hp) or 56kW (75hp) six-cylinder engine. The vehicle could carry 12 personnel and tow a light howitzer or anti-tank gun. The Blitz 3.6-36A was the four-wheel drive variant that could obtain superior tactical mobility across rugged terrain. Both the standard Blitz and Blitz-A were produced in six other main sub-variants: a field ambulance, a communications vehicle, a fire-engine, two types of fuel tanker and a rescue vehicle. Another variant was the

SdKfz 7 tractor

The Sdkfz 7 prime-mover featured a similar rear half-tracked arrangement as the Sdkfz 250, with five overlapping road wheels, two set forward and the rest mounted recessed behind them.

SdKfz 7 mittlerer Zugkraftwagen 8t

Crew: 11 personnel
Production: 1938–44
Weight: 9.2 tonnes (9 tons)
Dimensions: length: 6.85m (22ft 6in); width: 2.35m (7ft 9in); height: 2.62m (8.7ft)
Engine: 104kW (140hp) Maybach HL54
Speed: 50km/h (31mph)
Range: 250km (160 miles)

Prime mover towing payloads						
Vehicle	**Sdkfz 6**	**Sdkfz 7**	**RSO**	**Opel Blitz**	**Mercedes-Benz L3000**	**Krupp**
Carrying payload	1.5 tonnes (1.4 tons)	1.8 tonnes (1.7 tons)	1.5 tonnes (1.4 tons)	3 tonnes (2.9 tons)	2.65 tonnes (2.6 tons)	1.1 tonnes (1 ton)
Towing payload	5 tonnes (4.9 tons)	8 tonnes (7.9 tons)	2 tonnes (1.9 tons)	4 tonnes (3.9 tons)	3 tonnes (2.9 tons)	2 tonnes (1.9 tons)

lengthened 3.6-47 truck. A proportion of these vehicles were fitted in the field with either a 7.92mm (0.31in) MG 34 machine gun or a 20mm (0.79in) anti-aircraft gun. Widely used in all theatres of war, the Blitz proved a mechanically reliable vehicle even in demanding conditions, such as those in North Africa or the East.

Another cousin of the Blitz was the Opel 'Maultier' ('Mule'). This married the front part of the Blitz chassis to the tracks of the now-obsolete Panzer I tank to produce a half-tracked prime mover that could cope with the atrocious conditions on the Eastern Front. In total, the Germans produced over 82,300 Blitz trucks between 1937 and 1944.

Mittlere Lkw Opel Blitz 3000S
Crew: 1
Production: 1930–45
Weight: 3.29 tonnes (3 tons)
Dimensions: length: 6.02m (19ft 9in); width: 2.27m (7ft 5in); height: 2.18m (7ft 2in)
Engine: 55kW (74hp) Opel 6-cylinder petrol
Speed: 80km/h (50mph)
Range: 410km (255 miles)

Opel Blitz 3000
Half-heartedly camouflaged with natural vegetation, this image of a Opel Blitz truck, with canvas canopy erected, ably reveals the vehicle's high-sitting suspension

Mercedes-Benz L3000 medium truck

The Mercedes-Benz L3000 medium truck was a ubiquitous vehicle employed by the Wehrmacht in World War II. Built in three main variants – L3000, L3000A and L3000S – German factories produced 27,700 of these trucks in 1938–43.

The basic L3000 comprised a chassis with two wheeled axles, one located at the front and one (the drive axle) at the rear, on top of which was mounted a forward-located cab with a rear flatbed and shallow sides covering the rear section.

The standard L3000 weighed 3.85 tonnes (3.8 tons), was powered by a 55kW (74hp) Daimler-Benz OM 65/4 four-cylinder diesel engine and could reach a top speed of 64km/h (40 mph). Produced in 1938–39, the basic L3000 vehicle could transport a payload of up to 2.65 tonnes (2.6 tons).

The L3000S medium truck was a slightly narrower rear-drive variant that was powered by the up-rated 56kW (75hp) OM 65/4 engine. Produced in 1940–42, the vehicle weighed 3.69 tonnes (3.6 tons) and could obtain a top road speed of 69km/h (43mph). The final variant, the L3000A, was the four-wheel drive version produced in 1940–43. It weighed 4 tonnes (3.9 tons) and was powered by the 56kW (75hp) OM 65/4 engine.

The Wehrmacht employed the L3000 truck in all theatres of war to transport troops, equipment and stores. It was extensively employed in the open terrain of the North African campaign. Its drive crews, however, found that the

Mercedes-Benz L3000
This version of the truck has a high-sided cargo body for carrying equipment and supplies.

Mittlere Lkw Mercedes-Benz L3000
Crew: 1
Production: 1938–43
Weight: 3.85 tonnes (3.8 tons)
Dimensions: length: 6.25m (20ft 6in); width: 2.35m (7ft 8in); height: 2.6m (8ft 6in)
Engine: 55kW (74hp) Daimler-Benz OM 65/4 four-cylinder diesel engine
Speed: 64km/h (40mph)
Range: 410km (255 miles)

L 2-H 43 truck

This Protze L2 H 43 truck carries a spare wheel on the side of the driver's compartment, which remained very open thus affording him little protection from either the elements or from enemy fire.

Krupp Protze L2-H 43
Crew: 1, plus 5 passengers
Production: 1937–41
Weight: 2.6 tonnes (2.5 tons)
Dimensions: length: 5.1m (16ft 8in); width: 1.93m (6ft 4in); height: 1.96m (6ft 5in)
Engine: 45kW (60hp) Krupp M-304 6-cylinder petrol
Speed: 70km/h (45mph)
Range: 410km (255 miles)

extremely arduous conditions of the war in the East – rough, hard-baked ground or else deep, thick mud – placed excessive demands on the vehicle. Consequently, maintaining its tactical mobility proved challenging.

Krupp Protze L2-H 43 and H 143 medium trucks

The Krupp L2-H 43 and H 143 medium trucks were six-wheeled prime movers, with a single front axle and two rear axles. The front wheels had a semi-elliptic leaf-spring suspension arrangement and the independent rear wheels a horizontal coil-spring one. The low-located open driving position was located just behind the front wheel and engine, and there was a wooden rear flatbed section with low wooden sides positioned over the rear wheels.

The standard H 43 vehicle, produced in 1933–36, weighed 2.6 tonnes (2.5 tons) and was powered by either a 40kW (53hp) or 45kW (60hp) Krupp M-304 engine. From 1937 until 1941, production centred on the H 143 version, which incorporated minor changes to the positioning of the wheels and a modified Aphon gearbox.

Seven variants were produced in addition to the standard vehicle. These included: the Kfz. 70, which had seating mounted along the sides of the flatbed to transport 12 personnel; the Kfz. 19 telephone communications vehicle; the Kfz. 68 radio mast carrier; the Kfz. 81 ammunition transporter; the Kfz. 83 spotlight generator vehicle; and the Kfz.21 staff car variant. More than 7000 Krupp L2-H-143 trucks were produced between 1934 and 1941.

Chapter 2

Mortars, Artillery and Rockets

In addition to direct fire (principally delivered by tanks together with anti-tank, anti-aircraft and infantry guns), German offensive and defensive successes during the war owed much to the lethal and moral effects of indirect fire (IDF). The Germans employed three main types of IDF weapon in World War II: mortars, artillery pieces and rocket-launchers.

A mortar combined firepower with mobility to provide IDF at short range. Field artillery pieces (light, medium or heavy cannon or field howitzers/guns) delivered rounds across large distances for medium- or long-range effect. Rocket-launchers initiated into flight spin-stabilized rockets out to short and medium range. The most important examples of these weapons are detailed in this chapter.

MORTARS

This weapon was a high-elevation, smooth-bore device that fired a fin-stabilized bomb on a high plunging trajectory.

5cm (1.96in) leGW 36 light mortar

The 5cm (1.96in) leichte Granatwerfer 36 (leGW 36) light mortar was the second such weapon that the Germans

Opposite:
37mm (1.46in) anti-aircraft gun, Italian Front
A 3.7cm (1.46in) Flak gun with large splinter shield (to protect the crew) that has been mounted upon what appears to be a half-tracked, flat-loading prime-mover (possibly an SdKfz 7/2) deploys while soldiers stand nearby; Italian campaign 1943–44. Note the tiny holes in the barrel flash suppressor.

5cm Granatwerfer 36
A two-man crew prepare a 5cm (1.96in) GrW 36 light mortar for firing; the loader places a round taken from the nearby munitions box down the weapon's barrel after the firer had adjusted the mortar's elevation.

developed; it entered service in 1936, just as the army underwent massive expansion. The leGW 36 fired a small 0.9kg (2lb) charge to a maximum range of only 500m (1640ft). The small charge delivered meant that the exploding warhead had only a modest lethal effect. Despite these shortcomings, the leGW 36 still became a standard weapon within German infantry companies

and served in large numbers during the September 1939 Polish and May 1940 Western campaigns. In early 1941, however, the Germans decided to terminate production of the leGW 36. The Army high command came to the view that further manufacture of the weapon was an ineffective use of scarce resources: the mortar was both too complex and too expensive to manufacture given that it delivered a modest lethal effect with limited reach.

The leGW 36 was progressively withdrawn from German front-line service during 1942. However, it continued to be employed as a platoon- or company-level asset in German Army and Waffen-SS training or rear-area occupation units until the end of the war. In mid-1944, for example, each of the nine infantry companies in the SS Landstorm Nederland security regiment deployed a mortar platoon equipped with three leGW 36s; elements of the unit fired some of these weapons amid bitter defensive fighting around Arnhem in mid-September 1944.

8cm (3.14in) leGW 34 medium mortar

The 8cm (3.14in) Granatwerfer 34 (8cm/3.14in GrW 34) medium mortar was the first such weapon to be developed by the Germans, entering service in late 1934. Its first

5cm (1.96in) leichter Granatwerfer 36
Crew: 2
Production: 1936–45
Weight: 14kg (31lb)
Barrel length: 46.5cm (1ft 6in)
Range: 50m (164ft) minimum; 510m (1673ft) maximum
Rate of fire: 15–25rpm
Ammunition: 0.9kg (2lb) TNT filled

developmental work had started as far back as 1923. The mortar was an 81.4mm (3.2in)-calibre weapon, despite its designation as an 80mm (3.14in) mortar. It fired a decently sized 3.4kg (7.5lb) warhead out to a maximum distance of 2400m (7874ft). The mortar was sizable, with a barrel length of 1.14m (3ft 9in) and a weight of 62kg (137lb). This medium mortar was sturdy enough to survive the rigours of combat and proved a reliable weapon. It soon acquired a good reputation for accuracy, thanks to its RA-35 dial sight. This medium mortar remained the standard German infantry mortar throughout World War II, being employed in large numbers in every campaign in which the Wehrmacht fought. The weapon was usually operated by a three-man team.

In the late 1930s, the slightly modified 8cm (3.14in) GrW 34/1 was developed for employment from armoured vehicles, such as half-tracks and tractors. In addition, the kurz 8cm (3.14in) GrW 34 was a shortened version of the standard medium mortar that the Germans developed in 1940–41; it was just 75cm (29.5in) long and weighed 30kg (66lb). This weapon was issued principally to German paratrooper formations; as the war progressed, it was also fielded as a replacement for the phased-out 5cm (1.96in) leGW 36 light mortar.

SS with 8cm at Kursk
A Waffen-SS 8cm (3.14in) le GrW 36 mortar team shelter temporarily in a slightly sunken dirt road during the July 1943 'Citadel' offensive at Kursk; a properly-prepared weapons' pit would have offered both crew and weapon much greater protection than evident here.

8cm Granatwerfer 34
Side view of an 8cm (3.14in) leGrW 36 medium mortar, with bipod (left) and barrel and base plate (right); at the top of the bipod is the weapon's traversing gear and elevating mechanism.

8cm (3.14 in) leichter Granatwerfer 34
Crew: 2
Production: 193645
Weight: 62kg (136lb)
Barrel length: 1.14m (3ft 9in)
Range: 400m (440ft) minimum; 1200m (1310ft) maximum
Rate of fire: 15–25rpm
Ammunition: 3.5kg (7lb 11oz)

8cm Granatwerfer 42

A German crew train with the GrW 42 somewhere on the Eastern Front. The GrW 42 was an attempt to give German infantry units a close support weapon with greater hitting poweer than the mortars used in general service at the time.

12cm (4.72in) GrW 42 heavy mortar

The largest mortar to see operational service with the German armed forces was the 12cm (4.72in) Granatwerfer 42 (GrW 42) heavy mortar. The weapon was developed during 1942–43 in reaction to early encounters with the effective Soviet 122mm (4.8in) PM-38 regimental heavy mortar; in reality, the Germans designed a virtual copy of the Red Army weapon. The 12cm (4.72in) GrW 42 fired a hefty 15.8kg (35lb) warhead out to a maximum range of 6050m (19,850ft); both the lethal effect and range of the weapon was impressive given the typical capabilities of mortars of the day. In the hands of German personnel lucky enough to have this rare weapon, its potent lethality was widely appreciated. When movement was required, troops could quickly attach the mortar to a two-wheeled carriage that could be towed behind a vehicle. The weapon was expensive to produce, however, and was therefore produced in relatively limited numbers.

This limited production ensured that the weapon served with only a select few German formations. In August 1944, for example, a scheme was introduced where the 12cm (4.72in) GrW 42 would be issued to the heavy companies of select panzergrenadier and infantry regiments in place of medium mortars, but production shortfalls meant this was rarely executed. The 9th SS Panzer Division *Hohenstaufen* was an exception as it did receive 23 12cm (4.72in) GrW 42 mortars in summer 1944. The weapon was also issued in larger

12cm (4.72in) Granatwerfer 42
Crew: 2
Production: 1942–45
Weight: 280kg (620lb)
Barrel length: 1.8m (5ft 11in)
Range: 6050m (19,850ft) maximum
Rate of fire: 8–10rpm
Ammunition: 15.8kg (35lb)

Mortars compared			
Model	**5cm (1.96in) leGW 36**	**8cm (3.14in) GrW 34**	**12cm (4.72in) GrW 42**
Barrel life	25,000 rounds	20,000 rounds	3000 rounds
Muzzle velocity	174m/s (570ft/s)	223m/s (730ft/s)	283m/s (930ft/s)

numbers to independent units held at corps- or army-level. Further examples were issued to the regimental cannon companies of select panzergrenadier regiments in place of infantry guns such as the 15cm (5.9in) SiG 33.

FIELD ARTILLERY

Probably the most significant weapon type that the Germans used to deliver IDF during World War II was field artillery.

7.5cm (2.95in) leFK 18 light field gun

During the last year of the war, the German Army turned back to the employment of 75mm (2.95in) light field guns; these had been used in the early 1930s, but the Germans had then switched to the larger 105mm (4.15in)-calibre for this category of weapon. In 1930–31, the Germans married a Krupp-designed carriage with a Rheinmetall gun to create the 7.5cm (2.95in) leichte Feldkanone 18 (le FK 18). As with most German artillery weapons, it was either horse-drawn or moved by motorized vehicles.

This weapon remained in limited production throughout the 1939–45 war; however, because it was cheap and easy to produce, greater manufacturing resources were allocated to it in 1944–45. This weapon increasingly replaced infantry guns in the regimental cannon companies of German infantry divisions and SS grenadier formations during the war's last months. Weighing 1120kg (2464lb), this light field piece could fire a 5.8kg (13lb) shell out to a maximum range of 9425m (30,920ft).

Late on in the war, the Germans also concluded that it would be prudent to field a dual-purpose 75mm (2.95in) gun that could be employed in both the anti-tank and light field artillery roles. Consequently, during the winter

Feldkanon 18, camouflage net
This 7.5cm leFK 18 field gun has been emplaced in an open position adjacent to trees and bushes, and partially concealed with netting draped over the available foliage.

of 1944–45, the Germans introduced into service the improvised 7.5cm (2.95in) Feldkanone 7M58 and 7M59 dual-purpose field artillery/anti-tank guns. The former mounted the gun of the Pak 40 anti-tank weapon onto the chassis of the leFH 18/40 howitzer carriage.

In contrast, the 7M59 was essentially a slightly modified 7.5cm (2.95in) Pak 40 gun with increased elevation capability so that it could serve as an improvised artillery piece. These two models were only produced in small numbers, as the German war economy collapsed; they remained very rare assets on the battlefield in the war's final months.

10cm (3.94in) K 18 and K 18/40 heavy artillery gun

Another rare German artillery piece was the 10cm (3.94in) heavy cannon. This was produced in two variants: the schwere 10cm (3.94in) Kanone 18 and the K 18/40 (s 10cm/3.94in K 18 and K 18/40). Developed in 1926–29 by Rheinmetall and Krupp, the K 18 cannon entered service during 1933. This long-barrelled (L/52) gun was optimized to deliver very long-range counter-battery and interdiction fire out to an impressive maximum distance of 19,015m (62,385ft). The weapon was mounted on a standard two-wheeled split-trailed carriage; cannon that were moved by horses had aluminium wheels, whereas those moved by motor transport had solid rubber tyres.

Given its status as a specialized weapon, the Spreewerk firm in Berlin-Spandau only manufactured it in limited quantities in 1933–43; the total number in service peaked at just 762 in June 1941 and subsequently declined. A motorized cannon battery of four such guns often served in select German Army and SS motorized divisions.

Increasingly, however, the Germans transferred this heavy weapon to static emplacements within the numerous coastal artillery batteries established along the 'Atlantic Wall' fortifications along the coast of northwestern Europe.

The modified K 18/40 cannon was also developed by the firms of Krupp and Rheinmetall, this time in 1937–41. This design married an even longer-barrelled (L/60) gun, the 10cm (3.94in) K 40, to a modified version of the existing carriage of the K 18. This weapon developed a muzzle velocity of 905m/s (2970ft/s), eight per cent greater than its predecessor. This capability enabled the cannon to deliver its rounds out to an even greater maximum range of 21,150m (69,690ft), an 11 per cent increase on that obtained by the K 18. This version was produced in even smaller numbers than its predecessor.

10.5cm (4.13in) leFH 18 light field howitzer

In the late 1920s, the firm of Rheinmetall began developing the 10.5cm (4.13in) leichte Feldhaubitze 18 (leFH 18) light field howitzer. This design work led to the first prototypes being constructed in 1933, and the weapon entered general service in

1935. The howitzer weighed 1985kg (4377lb) and fired a 14.8kg (33lb) shell. The weapon discharged this shell at a muzzle velocity of 470m/s (1542ft/s), enabling it to reach an impressive maximum range of 10,675m (35,030ft).

In general production throughout the second half of the 1930s, the Germans fielded some 5200 leFH 18 howitzers during the September 1939 invasion of Poland, where it served as the standard light divisional howitzer. Production continued throughout the war, and it remained the standard weapon equipping the majority of divisional light artillery batteries.

Combat experience in Poland, the West, North Africa and the East in 1939–41 showed that although the howitzer was accurate and reliable, it was heavy to manoeuvre and its range was being outclassed by newer Allied weapons. In response during 1942, the Germans developed the modified leFH 18M. This howitzer sported a muzzle brake and redesigned recoil system

Coastal defences
Deployed in a seemingly open field position located close to the cliff edge overlooking the town of Arromanches in Normandy is a 10.5cm (4.13in) leFH 18 gun; this bucolic scene soon became replaced by one featuring the full fury of modern industrialized warfare.

to accommodate a more powerful propellant charge that increased the weapon's maximum range to 12,325m (40,440ft).

In combat, however, both the leFH 18 and 18M proved rather heavy and thus hard to manoeuvre. Consequently, the Germans improvised a lighter variant in 1942. They fitted the howitzer's barrel to the carriage of the 7.5cm (2.95in) Pak 40 anti-tank gun to produce the leFH 18/40. This marriage of convenience, however, only slightly reduced the howitzer's weight and thus only marginally improved its mobility.

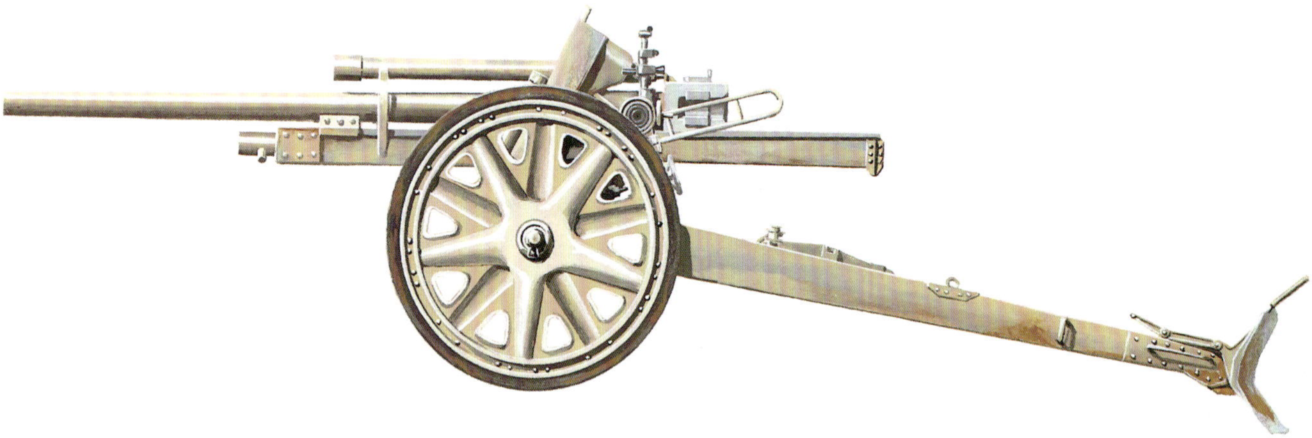

leFH 18

The leFH 18M (lower) was a modified version of the leFH 18 (top). The 18M is easily identifiable by its single-baffle barrel muzzle-break and by its redesigned recuperator and buffer assembly.

10.5cm (4.13in) leichte Feldhaubitze (leFH 18)

Crew: 6
Production: 1935–45
Weight: 2 tonnes (1.8 tons)
Length: 3.3m (10ft 11in)
Calibre: 105mm (4.13in)
Muzzle velocity: (AP) 540m/s (1770ft/s)
Range, HE: 12,325m (40,440ft)
Ammunition: Armour-piercing, high-explosive

Self-propelled leFH 18
With insufficient Wespes and Hummels available, new kinds of self-propelled artillery was always in demand. Here, a 10.5cm leFH 18 gun has bene mounted on a French Hotchkiss H39 chassis. The German 21st Panzer Division was particularly adept at this kind of improvisation.

10.5cm leFH 18 (sf) auf Geshützwagen 39H
Crew: 4
Production: 1944
Weight: 13.8 tonnes (12.5 tons)
Length: 4.7m (15ft 4in)
Calibre: 105mm (4.13in)
Muzzle velocity: (AP) 540m/s (1770ft/s)
Range, HE: 12,325m (40,440ft)
Ammunition: Armour-piercing, high-explosive

10.5cm (4.13in) leFH 43 light field howitzer

In 1943, the Army issued a demanding statement of requirement for the development of a new light field howitzer that paradoxically had both greater range than the well-proven leFH 18 yet was also more manoeuvrable. Only two German firms were prepared to take on this challenging requirement: Krupp and Skoda. The former's more conventional prototype lost out in trials to Skoda's novel design, which became designated the leichte Feldhaubitze 43 (leFH 43). At 2200kg (4851lb), this weapon was heavier than its predecessor, and its gun developed a significantly higher muzzle velocity of 610m/s (2002ft/s). The latter achievement, however, enabled the howitzer to reach the greater maximum range of 15,047m (49,370ft) – more than 4325m (15,520ft) further than the leFH 18. Skoda came up with an ingenious solution to solve the conundrum of how to make this heavier gun more manoeuvrable. Instead of a standard two-wheeled split-trail carriage, the gun was mounted on four outriggers. Two of these outriggers folded up underneath the barrel (to which they could be attached) during transportation. This mechanism reduced weight while allowing 360-degree traverse. However, in 1944–45, repeated Allied aerial attacks on German factories, railways lines and key industries (such as coal, steel and petro-chemicals) severely retarded production of the leFH 43. As a result, only a small number entered service in the last weeks of the war. Amid the chaos of Germany's last desperate defensive actions, the FH 43 made little discernible impact despite its technical qualities.

sFH 18 medium field howitzer
The standard 15cm (5.9in) sFH 18 was produced without a muzzle brake (unlike the later 18M variant); this medium howitzer was identifiable by its large recuperator mounted high above the weapon's barrel.

15cm (5.9in) schwere Feldhaubitze (sFH18)
Crew: 7
Production: 1933–45
Weight: 5512kg (12,154lb)
Length: 4.4m (14ft 5in)
Calibre: 150mm (5.9in)
Muzzle velocity: 520m/s (1706ft/s)
Rate of fire: 4rpm
Range: 13km (8 miles)
Ammunition: High-explosive or smoke

15cm (5.9in) sFH 18 medium field howitzer

In 1926–30, the firms of Krupp and Rheinmetall produced rival prototypes for the requirement for the army's future standard medium field howitzer. Elements of both designs were amalgamated, with the latter firm's gun being married to the former's carriage in 1933 to produce the 15cm (5.9in) schwere Feldhaubitze 18 (sFH 18); technically, despite its name, this was a 'medium' artillery weapon. Production of the gun commenced in late 1933 and it entered service in 1934. This piece remained the principal German medium field howitzer throughout World War II and was widely used in all campaigns. Four firms – Spreewerk, MAN, Doerres-Fuellner and Skoda – continued to produce the howitzer right up until the last weeks of the war.

The sFH 18 was of conventional design, with a typical split-trail carriage. The weapon weighed 5512kg (12,154lb) and could fire a 43.5kg (96lb) shell out to a maximum range of 13,250m (43,470ft), which was not particularly impressive for a weapon of this calibre. Many of these howitzers were horse-drawn; when in this mobility mode, the weapon's main carriage and rear limbers were transported separately from one another.

Other artillery batteries equipped with the weapon were motorized; these howitzers were moved in one section, with the four wheels being fitted with solid rubber tyres. In 1942, a modified version of this weapon was developed: the sFH 18M. This gun fired its round with a larger propellant charge to gain additional range. The increased charge, however, increased the rate of barrel erosion, so the design incorporated a replaceable chamber liner. In addition, the barrel featured a muzzle brake to reduce the stress placed on the carriage during firing.

Comparison of field artillery barrel lengths							
Model	**10.5cm (4.13in) leFH 18**	**10.5cm(4.13in) leFH 43**	**15cm (5.9in) sFH 18**	**17cm (6.7in) K 18in Mors.Laf**	**7.5cm (2.95in) FK 18**	**10cm (3.94in) FK 18**	**10cm 3.94in) FK 18/40**
Barrel length (in calibres – L/xx)	L/28	L/28	L/29.5	L/50	L/26	L/52	L/60

17cm (6.7in) K 18 in Mrs.Laf heavy artillery gun

The German army's heavy artillery pieces – typically of calibres of 17cm (6.7in) and above – generally served in independent heavy artillery batteries held at corps- and army-level. The most common heavy artillery piece deployed during the war was the 17cm (6.7in) Kanone 18 in Mörserlafette (17cm/6.7in K 18 in Mrs.Laf). This was a long-barrelled (L/50) long-range 172.5mm (6.79in) cannon designed for counter-battery work; that is, to destroy enemy artillery pieces. The gun was mounted in a 'mortar' style carriage that permitted an unusually high elevation of the barrel. The carriage was of advanced design that included a sophisticated dual-recoil system, in which – in addition to the barrel recoil – the whole platform also recoiled along its carriage rails. This arrangement astonishingly enabled a single soldier to quickly rotate the entire heavy 17,510kg (38,603lb) gun through 360 degrees.

The heavy gun entered limited production at the Hanomag firm in Hannover in 1941, but was never a common weapon. It proved to be the best German heavy gun of the war and could fire a 68kg (150lb) high-

21cm (8.27in) Mrs. 18
Crew: 10
Production: 1939–45
Weight: 16.7 tonnes (16.4 tons)
Length: 6.51m (21ft 4in)
Calibre: 210.9mm (8.3in)
Muzzle velocity: 565m/s (1854ft/s)
Rate of fire: 4rpm
Range: 16,700m (54,790ft)
Ammunition: High-explosive shell: 68kg (150lb)

21cm Mrs. 18 heavy howitzer
In addition to the 17cm (6.7in) K 18, the German Army also deployed the 21cm (8.27in) Mrs 18 heavy howitzer; in this artwork the large rectangular breach mechanism can be seen (coloured grey).

explosive round to an impressive range of 28,000m (91,860ft). The independent Wehrmacht units that deployed this gun were invariably motorized, with the cannon being transported in two separate loads.

However, for short tactical movements, vehicles could tow the entire weapon in a single section. Some eight Wehrmacht heavy battalions fielded the 17cm (6.7in) K 18 Mrs.Laf gun; in addition, one solitary Waffen-SS unit – the 101st SS Heavy Artillery Battalion (part of the elite I SS Panzer Corps) – fielded the weapon, having two batteries of four guns each.

ROCKET LAUNCHERS

The 1919 Treaty of Versailles banned Germany from developing heavy artillery, but did not restrict German rocket development because such weapons had not been widely employed in Europe. This oversight spurred the interwar German Army to develop rocket artillery.

15cm (5.9in) NbW 41 rocket-launcher

With the camouflage designation Nebelwerfer ('smoke projector'), the Germans covertly developed the first effective modern rocket systems. In 1940, the army accepted delivery of the 15cm (5.9in) Nebelwerfer 41 (15cm/5.9in NbW 41). The weapon comprised six 15.8cm (6.2in) launcher tubes fixed together in a circle that were fitted on twin pivots mounted upon a slightly modified 3.7cm (1.46in) Pak 35/36 anti-tank gun carriage. The weapon fired 150mm (5.9in) spin-stabilized rockets that carried both high-explosive and smoke warheads.

Panzerwerfer 42/SdKfz 4/1

The Panzerwerfer 42 weapons system mounted ten 15cm (5.9in) Nebelwerfer 41 rocket-launcher tubes (in two rows of five) on top of the rear of the Sdkfz 4/1 half-track, one of several designs of vehicles termed 'Maultier' ('Mule').

15cm (5.9in) Panzerwerfer 42 auf Sf (Sd Kfz 4/1)
Crew: 3
Production: 1941–45
Weight: 7.8 tonnes (7.1 tons)
Dimensions 6m (19ft 6in) x 2.2m (7ft 2in) x 2.5m (8ft in)
Engine: Opel 3.6l 6-cylinder
Armament: 15cm (5.9in) Nebelwerfer 41
Calibre: 158mm (6.22in)
Maximum range: 6.9km (4 miles)

Fully loaded with six rockets, the NbW 41 weighed 770kg (1694lb). The system could fire six 34kg (75lb) Wurfgrenate Spreng 41 high-explosive rockets out to a maximum range of 6900m (22,640ft). It took just ten seconds to fire a full salvo of six rockets. However, the two-man crew had to manually reload the device with new rockets; consequently, the weapon could fire only one six-rocket salvo every 100 seconds.

This launcher was not produced in large numbers and remained fairly uncommon. From 1941 onwards, the Germans allocated newly produced launchers to the elite rocket projector corps. The weapon first equipped independent rocket-launcher battalions and subsequently regiments and brigades. These weapons provided invaluable service substituting for the often inadequate levels of IDF being provided by the hard-pressed German artillery. Crucially, the launcher was highly mobile, which enabled the use of 'shoot and scoot' tactics. This mobility was essential to prevent high attrition rates at the hands of accurate Allied counter-battery artillery fire or devastating tactical air power sorties. As the war progressed, the Germans increasingly relied on their rocket-launchers as a more survivable asset than the less mobile and easier-to-locate artillery guns.

21cm (8.27in) NbW 42 rocket-launcher

In late 1942, the Germans introduced into service a heavier rocket-launcher, the 21cm (8.27in) Nebelwerfer 42 (21cm/8.27in NbW 42). When fully loaded, the weapon weighed 1100kg (2425lb) and had five, instead of six, barrels, in order to keep its overall weight down. Each 21cm (8.27in) Wurfgranate 42 (WGr 42) rocket weighed 113kg (249lb). The launcher could deliver these devices to a maximum range of 7850m (25,750ft).

Nebelwerfer 42

This artwork of a 21cm (8.27in) Nebelwerfer 42 heavy rocket-launcher reveals how basic the weapon was; it merely comprised five launcher tubes set upon a small vertical pedestal mount fitted to a simple split-trail carriage.

21cm (8.27in) Nebelwerfer 42
Crew: 3
Production: 1942–45
Weight: 1100kg (2425lb)
Length: 1.25m (4ft 1in)
Calibre: 21cm (8.27in)
Muzzle velocity: 320m/s (1050ft/s)
Range: 7850m (25,755ft)
Ammunition: High-explosive

30cm (12in) Nebelwerfer 42 rocket launcher

A small number of 28/32cm (11/12.6in) Nebelwerfer rocket-launchers were converted to take the new 30cm (12in) Wurfkorper 42 rockets, creating a weapon designated the 30cm Nebelwerfer 42. This one is being camouflaged in a farmer's field somewhere on the Eastern Front.

The NbW 42 could fire a full salvo in eight seconds. Like its smaller cousin, the rockets had to be manually reloaded by the crew, so the weapon could only fire three full salvoes in five minutes. When firing, the crew retired to the rear to avoid the back-blast and pulled on the electrical firing cable.

The 21cm (8.27in) NbW 42 was essentially an enlarged version of the NbW 41, and was mounted on an identical carriage. Hastily designed to rush it into production, it had a more conventional design. Unlike its predecessor, it delivered only high-explosive charges. Also like its smaller predecessor, the 21cm (8.27in) launcher delivered its rockets with reasonable accuracy thanks to some clever interwar development work. Historically, rockets had been inaccurate because the positioning of the propellant charge at the rear of the rocket produced in-flight instability. The German rocket-launchers, however, had the propellant charge at the front of the rocket, the back-blast of which was exhausted through Venturi devices at the rear of the rocket around the high-explosive charge. By angling these Venturis slightly, the Germans imparted spin to their rockets; the spin-stabilization thus created enhanced the weapon's accuracy and maximum obtainable range. The weapon was employed on all fronts during the last half of the war.

28/32cm (11/12.6in) NbW 41 rocket-launcher

The third major German rocket-launcher developed during the war was based on a substantially different firing arrangement than the other two. The 28/32cm (11–12.6in) Nebelwerfer 41 (28/32cm/11/12.6in NbW 41) was a dual-calibre weapon that could fire both the fin-stabilized 28cm (11in) and 32cm (12.6in) Wurfkorper rockets. Rather than circular barrels, however, this weapon featured a metal box-like frame mounting with three launching frames alongside one another on the top row, plus a further three below

Nebelwerfer production 1942–45				
Year	**1942**	**1943**	**1944**	**Jan–May 1945**
Launchers produced	3864	1706	3767	460

Rocket Launchers Compared				
Model	**Barrels**	**Barrel design**	**Velocity**	**In service**
15cm (5.9in) NbW 41	Six	Tubular	340m/s (1115ft/s)	1941–45
21cm (8.27in) NbW 42	Five	Tubular	320m/s (1050ft/s)	1942–45
28/32cm (11/12.6in) NbW 41	Six	Frame	145m/s (475ft/s)	1941–45

them on the bottom row. The frames as constructed fired the 32cm (12.6in) Wurfkorper Flamm incendiary rocket. The crew could also insert internal guide rails into the frame to enable the weapon to fire the 28cm (11in) Wurfkorper Spreng high-explosive rocket. This gave the device greater flexibility than the other major rocket-launcher designs. The launching frame was mounted on top of a two-wheeled trailer. The weapon was designed primarily as an anti-personnel device (with proven lethal and concussive effects), although the rockets also delivered good lethality against open or soft-skinned vehicles.

Some rocket-launcher battalions had one of their four subordinate batteries equipped with six 28/32cm (11/12.6in) launchers, so that 36 of the battalion's 144 total barrels were represented by this weapon. The combat experiences of these units, however, soon highlighted the major deficiencies that bedevilled the weapon. The system could develop only modest exit velocities for its

Nebelwerfer 41 on SdKfz 251 half-track

The Germans converted small numbers of Sdkfz 251 half-tracks to carry three angled 28/32cm (11/12.6in) Nebelwerfer 41 launch frames on each side of the vehicle; this system was termed 'Wurfrahmen 40' ('Launch Frame 40').

28/32cm (11/12.6in) Nebelwerfer 41
Crew: 3
Production: 1941–45
Weight: 1130kg (2490lb)
Length: 1.25m (4ft 1in)
Calibre: 28cm (11in); 32cm (12.6in)
Muzzle velocity: 320m/s (1050ft/s)
Range: 1925m (6315ft) (28cm/11in)
Ammunition: High-explosive

projectiles and thus could only deliver them over short ranges; it could fire the 28cm (11in) rocket out to just 1925m (6315ft) and the 32cm (12.6in) device slightly further, to 2200m (7220ft). These ranges were only a third of the distance that the 21cm (8.27in) rocket-launcher could obtain. In addition, the Worfkorper rockets proved less accurate than the Wurfgranate rockets delivered by the 15cm (5.9in) and 21cm (8.27in) rocket-launchers. As a result, production of the device remained limited.

ANTI-AIRCRAFT GUNS

To operate effectively in combat, German divisions had to counter with direct fire the various threats posed by enemy aircraft, AFVs and infantry. To deal with the threat posed by enemy ground-attack aeroplanes, German units deployed various direct-fire anti-aircraft guns, such as the famous 'Flak 88'. The Germans also sought to block the actions of enemy tanks and other AFVs by deploying well forward a range of direct-fire anti-tank guns, such as the legendary 8.8cm (3.4in) Pak 43.

Finally, in addition to the indirect fire of mortars and artillery, the Germans employed infantry guns to counter the threat posed by enemy infantry. Below we examine the most important examples of these weapons.

Flakpanzer 38(t)
The Flakpanzer 38(t) anti-aircraft AFV mounted the 2cm (0.79in) Flak 38 L/65 cannon onto the rear of the restructured Panzer 38(t) Model M tank chassis; during 1943–44 the Germans produced 142 of these vehicles.

Flakpanzer 38(t) Ausf M (SdKfz 140)
Crew: 5
Production: 1944–45
Weight: 9.8 tonnes (9.3 tons)
Dimensions: 4.61m (15ft 2in) x 2.13m (7ft) x 2.25m (7ft 5in)
Engine: 185kW (248hp) Praga AC 6-cylinder petrol
Speed: 42km/h (26mph)
Calibre: 20mm (0.79in)
Range: 2200m (7220ft)
Armament: 2cm (0.79in) Flak 38 L/65 cannon
Rate of fire: 120–180rpm

2cm (0.79in) Flugabwehrkanone [Flak] 38 L/65 light anti-aircraft gun

In the late 1930s, the Germans developed the 2cm (0.79in) Flak 38, an improved version of the existing ubiquitous 2cm (0.79in) Flak 30, already serving in very large numbers. This new 20mm (0.79in) weapon addressed two of the latter's main deficiencies: its slow sustained tactical rate-of-fire of 120rpm, and its ammunition feed problems. The Flak 38 could obtain a more satisfactory sustained rate-of-fire of 220rpm.

The weapon typically consisted of the long-barrelled gun, fitted with a well-sloped dual-angle splinter shield, mounted either on a fixed firing platform or a Sonderanhänger single-axle wheeled trailer. Under mass production in 1935–44, the Flak 38 became the Army's standard flak weapon during 1940 and supplanted the 2cm (0.79in) Flak 30, although it never ousted the latter completely. Employed by all branches of the Wehrmacht, it also appeared mounted on various fixed and mobile platforms. The combined total of Flak 30s and 38s in service peaked in March 1944 at 19,692 before declining sharply to 10,531 by February 1945.

2cm (0.79in) Flakvierling L/65 light anti-aircraft gun

In the late 1930s, Mauser developed for the Kriegsmarine the 2cm (0.79in) Flakvierling 38, a quadruple mounting of the existing Flak 38. The weapon proved so effective that it was subsequently produced for both the

2cm (0.79in) Flakvierling 38 auf Zugkraftwagen 8t (SdKfz 7/1)
Crew: 7
Production: 1934–45
Weight: 1.16 tonnes (1.06 tons)
Dimensions 6.55m (21ft 6in) x 2.4m (7ft 10in) x 3.2m (10ft 6in)
Engine: 104kW (140hp) Maybach HL62TUK 6-cylinder
Armament: Flakvierling 2cm (0.79in) quad guns
Range: 2200m (7220ft)
Rate of fire: 280–450rpm

Flakvierling 38
The Germans mounted the lethal quad 2cm (0.79in) Flakvierling 38 on several vehicles, including (as seen here) the 8-tonne (7.6-ton) Sdkfz 7 to create the Sdkfz 7/1 vehicle.

Flak Guns in Service		
Month	**Total Flak 30 and 38 in service**	**Total Flakvierling 38 in service**
January 1943	16,985	1062
January 1944	19,001	2602
January 1945	11,999	3806

Army and the Luftwaffe. The weapon consisted of two vertically paired 20mm (0.79in) cannon situated either side of a pedestal mounting. The sustained practical weight-of-fire of the four combined cannon – 880rpm – was astonishing. Low-flying enemy aircraft that entered the stream of bullets the weapon sent skywards stood a high probability of being downed; the weapon also proved excellent at area suppression in a ground role.

3.7cm (1.46in) Flak 36 and 37 L/57 light anti-aircraft guns

In the late 1930s, the Germans developed the 3.7cm (1.46in) Flak 36 anti-aircraft gun, an upgraded development of the existing 3.7cm (1.46in) Flak 18. The employment of 37mm (1.46in) flak cannon offered the Germans greater range and hitting power than the existing 20mm (0.79in) weapons. The weapon comprised the long-barrelled (L/57) cannon mounted onto either a fixed platform or the new lightweight single-axle wheeled Sonderanhänger 52 carriage. Mounted on the latter, the weapon was lighter than its wheeled predecessor, the 3.7cm (1.46in) Flak 18. Some other mobile examples of the weapon were alternatively mounted on the special four-wheeled trailer, the Sonderanhänger 104. The 3.7cm (1.46in) Flak 36 gun developed a

37mm Flak 36 Sdkfz 7/2

The sister variant of the Sdkfz 7/1 was this vehicle, the Sdkfz 7/2; this mounted the 3.7cm (1.46in) Flak 36 onto the 8-tonne (7.6-ton) half-track; the gun could be elevated from -8 degrees through to +85 degrees.

37mm Flak 36

Left-side view of a 3.7cm (1.46in) Flak 36 with its gun shield folded down; note the typical conical flash suppressor fitted to the barrel's muzzle, which featured tiny air holes too small to be seen here.

3.7cm (1.46in) Fliegerabwehrkanone (Flak 36)
Crew: 5
Production: 1936–45
Weight: 1.5 tonnes (1.4 tons)
Length: 3.5m (11ft)
Calibre: 37mm (1.46in)
Muzzle velocity: 820m/s (2690ft/s)
Range: 4800m (15,750ft)
Ammunition: High-explosive or smoke

sustained tactical rate-of-fire of 100rpm, and a theoretical maximum cyclic rate-of-fire of 160rpm. Subsequently, the 3.7cm (1.46in) Flak 37 variant was developed. This was identical to the Flak 36, except that it used the Zeiss Flakvisier 37 sighting mechanism rather than the Flakvisier 35 or 36 device used in the Flak 36 weapon; small numbers of the Flak 37 alternatively employed the Flakvisier 40 sighting device.

In 1942–44, the firms of Dürkropp, DWM and Skoda produced more than 4500 Flak 36 and Flak 37 weapons. Production was terminated in 1944, however, in favour of a more effective 37mm (1.46in) anti-aircraft weapon, the 3.7cm (1.46in) Flak 43, which had entered service late the previous year. The Flak 43 combined reduced weight with an increased rate-of-fire of 230–250 rpm delivered by a new gas-operated breech-loading mechanism, to produce a potent anti-aircraft gun. Production of this weapon continued until the war's last weeks, by which time around 1200 had been completed.

8.8cm (3.4in) Flak 18 L/56 heavy anti-aircraft gun

The 8.8cm (3.4in) Flak 18 L/56 was the first of several famous 'Flak 88' anti-aircraft gun models that served in the war. During the interwar years, Krupp collaborated with the Swedish firm Bofors; the Flak 18, which entered service in 1934, was the result. The design of the Flak 18 was highly innovative. The gun was mounted on a novel firing platform of cruciform design, with four outriggers that were extended so that the weapon could be fired. On top of the centre of this platform was a tall pedestal upon which the gun sat; this arrangement enabled the gun to be elevated right up to 85

Towed 88mm flak, Sdkfz 7
The Sdkfz 7 (this one with its canvas roof erected) was the principal prime-mover employed for the 8.8cm (3.4in) Flak 18; for movement the weapon was attached to two single-axle bogie assemblies, a process that took under three minutes.

8.8cm (3.4in) Fliegerabwehrkanone (Flak 18)
Crew: 10
Production: 1933–45
Weight: 4985kg (10,992lb)
Length: 5.791m (20ft)
Rate of fire: 15rpm
Calibre: 88mm (3.4in)
Range: 8000m (26,250ft) effective ceiling
Ammunition: QF 88 × 571mm R

degrees, a key performance enabler for an anti-aircraft weapon. The entire weapon weighed 4985kg (10,992lb).

Although the weapon was optimized for the anti-aircraft role (for which it fired high-explosive rounds), it also had a useful ground combat capability, firing armour-piercing shells. The gun fired its high-explosive round with a muzzle velocity of 820m/s (2691ft/s) and could obtain a maximum vertical range of 8000m (26,250ft). The weapon also fired armour-piercing rounds at a muzzle velocity of 795m/s (2609ft/s) to an incredible maximum range of 14,680m (48,160ft).

In the ground role, the gun could penetrate 88mm (3.4in) of vertical armour at a range of 2000m (6560ft). With a well-trained crew, the hand-loaded weapon could obtain a maximum rate of fire of 15rpm. The Flak 18 first saw combat in the Spanish Civil War in 1936 with the German Condor Legion, where four four-gun batteries were deployed. This combat experience revealed that the weapon was highly accurate and effective in its subsidiary role as a ground combat asset, either in the anti-tank or the artillery roles. Both this battle experience and the lessons from field exercises showed that the barrel of the Flak 18 wore out after firing a meagre 1000 rounds.

The slightly modified 8.8cm (3.4in) Flak 36 went into production in the late 1930s. This addressed the barrel wear issue by incorporating a three-part barrel assembly that included easily replaceable barrel liners. The following year, the further modified Flak 37 variant entered service. This incorporated enhanced fire control equipment. Produced in large numbers, the Flak 18, 36 and 37 became the standard German heavy anti-aircraft weapons of World War II.

In September 1942, some 5184 Flak 18 and Flak 36 weapons were in German service. With production continuing through to almost the end of the war, the number of Flak 18, 36 and 37s in service peaked at 10,704 in

August 1944; by February 1945, this total had declined to 8769 weapons.

8.8cm (3.4in) Flak 41 L/74 heavy anti-aircraft gun

In early 1942, the Germans introduced into service a fundamentally new 88mm (3.4in) anti-aircraft gun, the 8.8cm (3.4in) Flak 41. Early teething problems, however, meant the device only got to the front in reasonable numbers in early 1943. This Rheinmetall-developed weapon was specifically designed to deliver a dual-role capability; thus it possessed genuine anti-tank stopping power. All things considered, this was a superior weapon to its predecessor Flak 88s. The gun, mounted within a three-sided shield, was the extremely long-barrelled L/74 model 88mm (3.4in) cannon. This potent gun fired its rounds at a muzzle velocity of 980m/s (3216ft/s). When delivering armour-piercing rounds, the weapon could penetrate 132mm (5.2in) of vertical armour at a range of 2000m (6560ft). Its maximum theoretical range in the ground role was an astonishing 20,000m (65,620ft).

During the 1941 German conquest of Greece, a Flak 41 of the SS *Leibstandarte* knocked out a British tank at a range of

88 Flak 18, Kursk Offensive
A 8.8cm (3.4in) Flak 18 gun deployed in an open field during the July 1943 'Citadel' offensive at Kursk; the large recuperator above the barrel is very apparent in this image.

6000m (19,685ft). Similarly, in the July 1944 British Goodwood offensive, a battery of Flak 41s destroyed 35 British tanks within a 48-hour period. When engaging enemy aircraft, the weapon could fire its high-explosive rounds out to a maximum vertical range of 14,690m (48,105ft). When its penetration capabilities and range were combined with the weapon's incredible rate of fire of 20–25rpm, the Flak 41 was a devastating weapon. The weapon could be made mobile by mounting it on the Sonderanhänger 202 four-wheeled double-tyred trailer.

The Flak 41 also featured enhanced survivability in that its design was less tall than its predecessors, reducing the size of target it presented to the enemy; it was just 2.36m (7ft 9in) in height. The high silhouette of former Flak 88 models was their most significant drawback in the ground combat role. The device also weighed a considerable 7800kg (17,199lb). This powerful weapon, however, was only produced in limited numbers so as not to disrupt the continuing large-scale production of its three predecessor Flak 88 models. In 1942, German factories produced 42 Flak 41s, with 122 produced during 1943.

Most of these weapons were deployed to the heavy batteries of the artillery battalions within a few elite panzer and Panzergrenadier divisions, or else to independent flak units. In 1944, some 290 Flak 41s were produced, together with a further 96 in early 1945. The vast majority of these weapons were issued to some of the hundreds of Army and Luftwaffe flak batteries deployed within the Reich itself in an attempt to disrupt the daily long-range Allied bombing attacks that were inflicting so much destruction upon the German war economy and transportation systems.

12.8cm (5in) Flak 40 L/61 heavy anti-aircraft gun

Rheinmetall began developing an 128mm (5in) heavy anti-aircraft gun in

Flak 41

This 8.8cm (3.4in) Flak 41 has been set at a high elevation, way above the top of the splinter shield. It has been mounted on its two single-axle limbers and its outriggers are visible extending between the bogie assemblies.

8.8cm (3.4in) Fliegerabwehrkanone (Flak 41)

Crew: 10
Production: 1943–45
Weight: 7840kg (17,284lb)
Length: 651.2m (2135ft)
Rate of fire: 20–25rpm
Calibre: 88mm (3.4in)
Range: 20,000m (65,620ft)
Ammunition: QF 88 × 571mm

Flak 40

Sitting on a substantial four-limbered H-shaped metal static carriage, the sheer size of the monstrous 12.8cm (5in) Flak 41 gun is evident here; note the wide girth of the pedestal mount upon which the weapon sits.

12.8cm (5in) FlaK 40
Crew: 10
Production: 1942–45
Weight: 4828kg (10,644lb)
Length: 7.835m (25ft 8in)
Rate of fire: 12–14rpm
Calibre: 128mm (5in)
Range: 14,800m (48,460ft)
Ammunition: 128 x 958mm

1936; at the same time, Krupp worked on two similar 150mm (5.9in) gun projects. In 1937, the first 12.8cm (5in) Flak 40 L/61 prototypes underwent testing; subsequently, Rheinmetall produced six examples while the prototype 150mm (5.9in) guns were being tested.

These first Flak 40 weapons were mobile and carried in one load on a special transporter: the Sonderanhänger 220 four-wheeled trailer. The protracted trials held to assess the capabilities of the prototype 150mm (5.9in) guns demonstrated that these designs were not superior to the Flak 40, and these projects were cancelled in early 1940. Consequently, Krupp produced 16 Flak 40s in 1942, but these were static rather than mobile weapons. Production of the static version of the Flak 40 increased thereafter, with 160 being built in 1943, 380 during 1944 and 50 in early 1945. The number of Flak 40s in service peaked in January 1945 at 570 pieces. Most of these guns served in some of the hundreds of Army and Luftwaffe heavy anti-aircraft batteries that defended the Reich from Allied bombing attacks.

The Flak 40 was, at 61 calibres, a long-barrelled 128mm (5in) cannon that could be elevated up to 87 degrees. The gun sat on a pedestal mount fitted upon a heavy four-limbed ground mounting. Firing its high-explosive rounds at a muzzle velocity of 880m/s (2887ft/s), the gun could obtain an impressive maximum vertical ceiling of 14,800m (48,460ft).

Capable of traversing through 360 degrees, the gun could deliver 12–14rpm. The weapon was bulky and heavy, weighing 4828kg (10,644lb). The Germans introduced minor adjustments to the design in 1944, which led to these vehicles being redesignated the Flak 40/1 and 40/2.

In the war's last weeks, many of these guns were used in an improvised ground role as Allied armoured spearheads raced deep into the heart of the Reich.

ANTI-TANK GUNS

The German deployed some of the best and most effective anti-tank guns of the war, many of which were also used to equip tanks.

3.7cm (1.46in) Pak 35/36 light anti-tank gun

The Germans introduced into service the motor-towed 3.7cm (1.46in) Flak 35/36 (1.46in) L/45 anti-tank gun in 1935; it replaced the similar horse-drawn weapon developed in the late 1920s. The weapon was small, low-silhouetted and readily manoeuvrable. Fitted to a steeply sloped splinter shield, the 45-calibre cannon fired its Pak PzGr anti-tank round at a muzzle velocity of 762m/s (2500ft/s) out to a maximum range of 4025m (13,205ft). At a combat range of 500m (1640ft), the gun could penetrate 48mm (1.9in) of vertical armour; this was sufficient

Pak 36

The large size of the tyred solid-form wheels are evident on this artwork of the 3.7cm (1.46in) Pak 36; notice as well the large spades at the rear of the trailer.

3.7cm (1.46in) Panzerabwehrkanone (Pak 36)

Crew: 3
Production: 1933–42
Weight: 0.43 tonnes (0.39 tons)
Length: 1.67m (5ft 5in)
Calibre: 37mm (1.46in)
Muzzle velocity: 762m/s (2500ft/s)
Range: 600m (1970ft)
Ammunition: Armour-piercing

to penetrate most enemy tanks from any angle. The gun was mass-produced in 1935–39 and became the standard German divisional anti-tank asset. Before the start of the September 1939 Polish campaign, the Army fielded 11,200 Flak 35/36 guns. During this campaign, the Pak 35/36 proved adequate to deal with the Polish Army's numerous light tanks and tankettes.

The 37mm (1.46in) gun's tactical limitations were exposed in the May–June 1940 German invasion of France: the weapon could not penetrate the 60mm (2.4in)-thick armour of the French Char B and British Matilda heavy tanks.

Subsequently, the Germans began developing a 50mm (1.96in) anti-tank gun, but in the meantime the 37mm (1.46in) weapon remained the standard anti-tank asset, with 14,458 in service in May 1941. To temporarily extend the weapon's service life, the Germans developed the higher-performance 3.7cm (1.46in) Panzergrenate 40 (PzGr 40) tungsten-carbide round; this could penetrate 65mm (2.6in) of vertical armour at 500m (1640ft). However, growing shortages of tungsten due to the Allied naval blockade meant that the round could be produced in only limited quantities.

Subsequently, the bitter fighting on the Eastern Front during the second half of 1941 again highlighted the growing tactical obsolescence of the 37mm (1.46in) gun. Even at extremely close range, the Pak 35/36 could not penetrate the frontal armour of the Soviet T-34 and KV-1 tanks. To again temporarily extend the weapon's service life, the Germans introduced the 3.7cm (1.46in) Stielgrenate 41 fin-stabilized hollow-charge HEAT (high-explosive anti-tank) bomb; this fitted over the muzzle of the gun and was fired by means of a blank cartridge in the breech. Although it possessed a range of just 200m (656ft), the Stielgrenate 39 increased the penetration of the Pak 35/36 to 180mm (7.2in). Despite these measures, the Germans began withdrawing the obsolescent 37mm (1.46in) gun from front-line service in 1942–43.

5cm (1.96in) Pak 38 L/60 medium anti-tank gun

Even before the deficiencies of the 3.7cm (1.46in) Pak 35/36 anti-tank gun had been exposed, Rheinmetall-

Anti-Tank Guns Compared

	Type	Penetration of vertical armour at 200m (656ft)	Muzzle velocity	Weight
3.7cm (1.46in) Pak PzGr	Solid-shot anti-tank	56mm (2.2in)	762m/s (2500ft/s)	0.68kg (1.5lb)
3.7cm (1.46in) PzGr 40	Tungsten carbide anti-tank	72mm (2.8in)	1030m/s (3392ft/s)	0.34kg (0.75lb)
Stielgrenate 41	Fin-stabilized hollow-charge HEAT (high-explosive anti-tank) bomb	180mm (7in)	110m/s (361ft/s)	8.5kg (19lb)

Borsig had begun developing a more potent replacement weapon, the 5cm (1.96in) Pak 38. This entered service in autumn 1940. The 50mm (1.96in) gun, fitted to a large splinter shield, was relatively light and mounted on an easily manoeuvrable single-axle split-trail carriage. Like many German anti-tank guns, when being moved tactically it was towed by a Sdkfz 251 half-track.

This long-barrelled (L/60) 50mm (1.96in) cannon, which was fitted with a muzzle brake, could fire the standard Panzergrenate 38 armour-piercing round out to a maximum range of 2650m (8700ft), and could penetrate 61mm (2.4in) of vertical armour at 1000m (3280ft).

Firing the rare Panzergrenate 40 tungsten-carbide round, the gun could penetrate 84mm (3.3in) of vertical armour at 1000m (3280ft); this round was the only German anti-tank weapon that at that time was able to penetrate the Soviet T-34 tank frontally at normal combat ranges. In the hands of a well-trained crew, the 50mm (1.96in) weapon could also deliver an impressive rate-of-fire of 12–15rpm.

During the summer 1941 Axis invasion of the Soviet Union, small numbers of 5cm (1.96in) Pak 38 guns equipped a single platoon within the anti-tank companies of elite Army and Waffen-SS motorized divisions. Subsequently, the 50mm (1.96in) gun remained in large-scale production in 1942–43, with 8500 examples produced in this period. It became the standard anti-tank weapon of the German Army, gradually replacing the 3.7cm (1.46in) Pak 35/36 in front-line service.

The gun served with the Afrika Korps in the North African campaign of 1942–43, as well as on the Eastern Front and the Western Front in late 1944. In 1944, however, the now increasingly obsolete gun was

Pak 38

Left-side view of 5cm (1.96in) Pak 38; both its split carriages had a spade device with a kinked manoeuvring handle.

5cm (1.96.in) Panzerabwehrkanone (Pak 38)

Crew: 3
Production: 1940–43
Weight: 1.2 tonnes (1.13 tons)
Length: 3.2m (10ft 5in)
Calibre: 50mm (1.96in)
Muzzle velocity: (AP) 835m/s (2900ft/s)
Range: AP: 1800m (5905ft), HE: 2.6km (1.6 miles)
Ammunition: Armour-piercing, high-explosive

Pak 40

This early-production 7.5cm (2.95in) Pak 40 sports the initial design single-baffle muzzle-brake as well as spoked steel wheels; later examples featured one of two slightly modified muzzle-brake designs.

7.5cm (2.95in) Panzerabwehrkanone (Pak 40)
Crew: 6
Production: 1942–45
Weight: 1.5 tonnes (1.37 tons)
Length: 3.7m (12ft 1in)
Calibre: 75mm (2.95in)
Rate of fire: 14rpm
Muzzle velocity: (AP): 933m/s (3061ft/s)
Range: AP: 2000m (6560ft), HE: 7.5km (5 miles)
Ammunition: Armour-piercing, high-explosive

withdrawn to second-line occupation duty roles. In total, Rheinmetall-Borsig manufactured 9504 Pak 38 guns.

7.5cm (2.95in) Pak 40 L/46 heavy anti-tank gun

Developmental work on a new 75mm (2.95in) anti-tank gun had begun in 1939 even as work on the 5cm (1.96in) Pak 38 unfolded. To speed up development, the Germans designed the new 75mm (2.95in) weapon as a scaled-up copy of the Pak 38. The first examples of this new weapon, the 7.5cm (2.95in) Pak 40 L/46, were rushed to the Eastern Front just as the Soviet winter 1941–42 counter-offensive erupted. The troops of the hard-pressed and out-flanked German defensive hedgehogs, like those at Velikiye Luki, were relieved when they discovered the new weapon's potent tank-killing powers; it could despatch even the feared T-34 and KV-1 tanks.

The 7.5cm (2.95in) Pak 40 was a large but low-silhouetted and easy-to-conceal weapon that resembled its smaller progenitor, the Pak 38. The long-barrelled (L/46) cannon was fitted to a three-sided angular splinter

5cm (1.96in) Pak 38 and 7.5cm (2.95in) Pak 40 compared

	Barrel life	Elevation	RSO
5cm (1.96in) Pak 38	4000–5000	-8 to +27	65
7.5cm (2.95in) Pak 40	6000	–5 to +22	65

PaK 40, Kursk Offensive
The crew of a very well foliage-camouflaged 7.5cm 2.95in) Pak 40 place a new round into the weapon's open breech at some point during the German July 1943 'Citadel' offensive at Kursk.

shield; this shield was one of the key features that distinguished the weapon from its visually similar forebear. The gun was mounted on top of a standard-design single-axled, pneumatic-tyred, split-trail carriage.

This effective weapon proved relatively cheap to produce. With full-scale mass-production underway in 1942–44, the weapon gradually replaced the Pak 38 as the standard anti-tank asset of German divisions in 1943–44. Many examples of the gun were also installed in the bunkers of the Wehrmacht and Kriegsmarine coastal artillery batteries that formed the Atlantic Wall fortifications along the western coast of Nazi-occupied France.

The Pak 40 delivered an impressive ballistic performance. Firing the 7.5cm (2.95in) Panzergrenate Patrone 39 armour-piercing, capped, ballistic-capped (APCBC) round at a muzzle velocity of 792m/s (2598ft/s), the gun could penetrate at 500m (1640ft) some 135mm (5.3in) and 106mm (4.2in) of vertical and 30-degree sloped armour, respectively. Firing the Patrone 40 tungsten-core round at a muzzle velocity of 933m/s (3061ft/s), the gun could penetrate at 500m (1640ft) some 154mm (6in) and 115mm (4.5in) of vertical and 30-degree sloped armour, respectively.

The Pak 40 served on all fronts right through until the war's end in May 1945.

PAK 40 TANK KILLER

One example of this gun's tank-killing prowess occurred on 8 August 1944 during the Canadian Totalize offensive in Normandy. That day near the Chateau du Fosse, 12th SS Panzer Division panzergrenadiers, reinforced by two well-camouflaged Pak 40s, were deployed in the adjacent wood. A 1st Polish Armoured Division Sherman tank squadron unknowingly moved towards them across open corn fields. Within 35 minutes, a hail of Pak 40 fire had knocked out 18 Polish tanks, although their return fire destroyed one of the German guns.

8.8cm (3.4in) Pak 43 L/71 anti-tank gun

In 1943, the firm of Krupp designed a technologically advanced and highly potent successor to the 7.5cm (2.95in) Flak 40, designated the 8.8cm (3.4in) Pak 43 L/71. This weapon is widely regarded as the most effective anti-tank gun of World War II. The Pak 43 mounted an extremely long 71-calibre barrel fitted with a muzzle brake. This lethal cannon fired the newly designed 10.4kg (23lb) PzGr 39/43 armour-piercing round at an astonishing muzzle velocity of 1000m/s (3300ft/s). Firing this round over the close tactical range of 500m (1640ft), the Pak 43 could penetrate 207mm (8.1in) of vertical

Panzerabwehrkanone 43
Crew: 5
Production: 1943–45
Weight: 3700kg (8160lb)
Length: 9.2m (30ft 2in)
Calibre: 88mm (3.4in)
Muzzle velocity: 1000m/s (3300ft/s)
Rate of fire: 6–10rpm
Range: 15,150m (49,700ft)
Ammunition: High-explosive or smoke

PaK 43
Side and rear views of the fearsome Pak 43. The gun's large rectangular breech can be clearly seen in the centre of the splinter shield in the rear-view, as can the two carriage-rear spades.

PaK 43 Compared				
	Weight (complete)	Elevation	Barrel life (rounds)	Manufacturers
Pak 43	4750kg (10,472lb)	–8 o to +40 °	1200–2000	Krupp; Henschel; Weserhütte
Pak 43/41	4380kg (9656lb)	–5 o to +38 °	1200–2000	Rheinmetall-Borsig

armour and 182mm (7.1in) of 30-degree sloped armour; this was sufficient to destroy any extant Allied AFV. Even at the longer-than-normal tactical range of 2000m (6560ft), the gun could penetrate 159mm (6.3in)-thick vertical armour. The weapon's maximum effective range was an equally impressive 15,150m (49,700ft).

The gun also featured an advanced semi-automatic breech-loader and electrical firing circuit that allowed it to obtain a rapid rate of fire of 6–10rpm. In 1944, the weapon gave the Germans the capability to destroy the new Soviet JS-1 and JS-2 heavy tanks, which had been impervious to the fire of the 7.5cm (2.95in) Pak 40. The sheer velocity of the rounds delivered, plus the high pressures within the barrel, meant that barrel wear was extreme; replacement was sometimes needed after a mere 1200 rounds fired. Later Pak 43 barrels were produced in two sections to ease replacement of worn sections.

The long gun was mounted on a low cruciform platform with a well-sloped splinter-shield to protect the crew. At just 2.05m (6ft 9in) tall, the weapon presented a lower silhouette than did the dual-role 8.8cm (3.4in) Flak, giving it superior survivability. Weighing 3700kg (8160lb), the Pak 43 was also easier to manoeuvre tactically than the Flak 88. The Pak 43 was a finely engineered weapon that was costly and consumptive in time and resources to construct.

With the field army clamouring for this superb tank-killer, production always lagged behind demand, with just 2100 Pak 43s being completed. Although allocated to just a handful of corps- and army-level independent anti-tank battalions, the Pak 43 made its mark on the tactical battlefield despite its small numbers. During the 17 July 1944 Anglo-Canadian Goodwood offensive southeast of Caen, for example, three Pak 43 guns deployed near Cagny knocked out 23 British tanks in under 30 minutes.

8.8cm (3.4in) Pak 43/41 L/71 heavy anti-tank gun

In 1944, the Army's clamour for the lethal 8.8cm (3.4in) L/71 anti-tank gun could not be met by supply; the Pak 43 was a complex and resource-intensive design to construct, particularly its sophisticated cruciform carriage. But with the Army desperate to get as many 88mm (3.4in) L/71 barrels to the front line as swiftly as possible, Rheinmetall-Borsig were tasked with urgently designing a new improvised easy-to-construct weapon. They designed a device that married a modified example of the existing L/71 barrel (with simpler sights and breech mechanism) to three elements: a new three-sided splinter shield; the split-trail carriage of the 10.5cm (4.13in) leFH 18 light field howitzer; and the wheeled single-axle of the 15cm (5.9in) sFH 18 heavy field howitzer. The latter two elements were in mass production and then available in large numbers. The resulting improvised mongrel weapon was designated the 8.8cm (3.4in) Pak 43/41.

The speed with which the weapon was designed, and its expedient nature, resulted in several tactical handicaps. The Pak 43/41 was significantly taller than the Pak 43 and thus presented more of a target at which the enemy could aim. Secondly, the gun could traverse through only 56 degrees, compared with the 360-degree capability of the Pak 43. A heavy gun provided with just one (not two) axles, moreover, the Pak 43/41 was also less tactically manoeuvrable than the Pak 43.

These deficiencies aside, the 8.8cm (3.4in) Pak 43/41 fired the same ammunition as the Pak 43 and delivered the same outstanding ballistic capability. In getting greater numbers of this potent weapon to the front line in whatever form, the Germans had made a sensible strategic decision. The Pak 43/41 joined its sister weapon in army-level independent anti-tank battalions in the last 14 months of the war.

INFANTRY GUNS

During the interwar period, the German Army developed a new genre of direct-fire weapon termed the infantry gun (Infanteriegeschütz). This was a lightweight, mobile direct-fire-support weapon operated by infantry that could obtain ranges beyond that of mortars.

7.5cm (2.95in) leIG 18 light infantry gun

In 1927–33, the Germans developed two infantry gun designs: the 7.5cm (2.95in) leIG 18 light weapon, and the 15cm (5.9in) SiG33 heavy device; both were to equip the newly forming cannon companies within an infantry regiment.

75cm (2.95in) leichtes Infanteriegeschütz 18
Crew: 3–5
Production: 1932–45
Weight: 405kg (893lb)
Barrel length: 88cm (34in)
Calibre: 75mm (2.95in)
Muzzle velocity: 210m/s (690ft/s)
Rate of fire: 8–12rpm
Range: 3550m (11,647ft)
Ammunition: High-explosive or smoke

leIG 18
The extremely short barrel of the leIG 18 is evident here – it barely projects much in front of the wide three-sectioned well-sloped splinter shield.

7.5cm leIG 18, Stalingrad
The five-man crew of a 7.5cm (2.95in) leIG 18 gun man-handle it in wasteland during the Battle of Stalingrad; note its unusual single-piece wide carriage, evident in the foreground.

The Rheinmetall-designed 7.5cm (2.95in) leIG 18 entered service in 1932. It came in two variants: the first, with spoked wheels, was designed to be horse-drawn; the second, with filled-in wheels that sported pneumatic tyres, was moved by motor vehicles. As production unfolded, a larger proportion of manufacturing capacity was given to the latter variant. The horse-drawn variant weighed 405kg (893lb) and the motor-drawn one 515kg (1135lb). The device mounted a very short-barrelled (L/11.8) 75mm (2.95in) gun that delivered a 6kg (13lb) round out to a maximum range of 3550m (11,647ft).

In 1944, a new infantry gun – the 7.5cm (2.95in) leIG 37 – began to supplement the leIG 18 in front-line German units. This new weapon married the barrel of the prototype leIG 42 infantry gun with existing gun carriages of the obsolescent 3.7cm (1.46in) Flak 35/36 anti-tank gun, which were then available for conversion in sizable numbers.

15cm (5.9in) SIG 33 heavy infantry gun

In 1927–34, the German Army developed a second, heavy, infantry gun design: the 15cm (5.9in) sIG 33. Limited production and field testing occurred in 1935–37; larger-scale sIG 33 production commenced in 1938. The weapons manufactured were primarily allocated to the cannon companies found within the German infantry regiment. From 1940 onwards, two 15cm (5.9in) sIG 33 guns were allocated to the heavy infantry gun platoon found in Waffen-SS Panzergrenadier and infantry regiments.

An infantry gun was supposed to be a mobile and manoeuvrable asset that front-line troops could position well forward to provide intimate direct-fire support, particularly if artillery indirect fire was unavailable. At 1700kg (3750lb), however, the sIG 33 was heavy – and also bulky – for such a forward-deployed role. That said, infantry guns could also function as Ersatz indirect-fire artillery pieces in extremis.

The short-barrelled (L/11.4) 150mm (5.9in) cannon delivered a high-explosive round out to a maximum range of 4700m (15,420ft); it could also fire smoke, hollow-charge, and HEAT sticky bomb (Stielgrenate 42) rounds. The gun was mounted on a non-split, single-axled, wheeled carriage. This platform came in two forms: an all-steel version, termed the Model A, and the Model B, which combined steel with lightweight metal alloys. The weapon could be motor-towed or horse-drawn.

15cm (5.9in) Schwere Infanterie Geschütz (sIG 33)
Crew: 5
Production: 1936–45
Weight: 1700kg (3750lb)
Length: 1.64m (64.57in)
Calibre: 150mm (5.9in)
Muzzle velocity: 241m/s (783ft/s)
Rate of fire: 2–3rpm
Range: 4700m (15,420ft)
Ammunition: High-explosive or smoke

sIG 33 heavy gun
The large manoeuvring handle of the sIG 33 positioned on the top of the non-split carriage rear spade is clearly evident in this artwork, as are the early-production spoked wooden wheels.

CHAPTER 3

Infantry Equipment

The basic weapon of the German Army during World War II was the personal small arm that the infantry soldier within the section carried. The basic weapon was the rifle or its shortened variant the carbine, often preferred by armoured, cavalry or artillery personnel. Other personnel carried a sub-machine gun instead, such as the MP 38 or 40.

As the war progressed, technological developments provided the infantry with new, more potent, small arms, such as the rapid automatic-fire MP 43/StG 44 assault rifle. In the same period, infantrymen received new personal anti-tank weapons such as the Panzerfaust and Panzerschreck. Finally, in addition to these personal weapons, the nine soldiers in a German section existed to preserve and deliver the fire-effect of its principal weapon, the machine gun; typically either the MG 34 or MG 42.

PISTOLS

Parabellum P'08 (Luger)

The Parabellum pistol's ancestry reaches back into the 19th century, but it is probably defined by the Pistole '08, known after its designer, Georg Luger. The transition from the 7.65mm calibre of Luger's early guns to 9mm Parabellum secured the popularity of

Opposite:
Tobruk
A German Army officer (foreground) and soldier grab some rest during operations at Tobruk in North Africa. The officer carries an MP 40 sub-machine gun and a stick grenade (*Stielhandgrenate*).

Parabellum P'08
This cutaway of a Luger P'08 shows a round in the chamber, sitting on the firing pin behind it while behind that the firing pin spring runs horizontally back towards the circular front toggle link.

the Pistole '08, and it was adopted by both the German navy and army. More than 2.5 million were subsequently produced between 1908 and 1945. Ironically, its visually distinctive toggle-lock mechanism was both a virtue and a problem – a virtue in that it worked well, but a problem in that it only worked well if kept clean, something far from easy in combat conditions. Nevertheless, it fired accurately, was comfortable for the firer to hold and, even when simplified for wartime manufacture, exuded quality.

Parabellum P'08
Date: 1908
Calibre: 9mm Parabellum
Operation: Short recoil
Weight: 0.87kg (1.92lb)
Length: 233mm (8.75in)
Muzzle velocity: 380mps (1247fps)
Feed/magazine capacity: 8-round detachable box
Range: 30m (98ft)

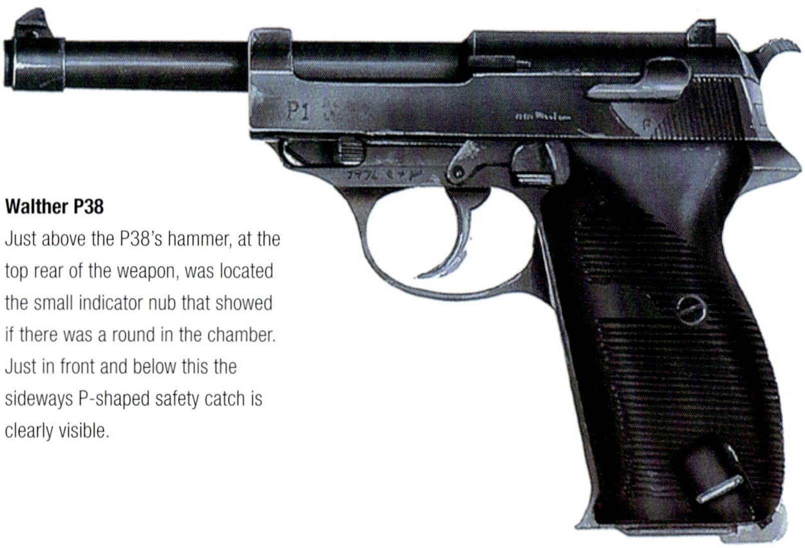

Walther P38
Just above the P38's hammer, at the top rear of the weapon, was located the small indicator nub that showed if there was a round in the chamber. Just in front and below this the sideways P-shaped safety catch is clearly visible.

Walther P38

Part of Germany's armaments rationalisation and expansion in the 1930s was the request for a new service pistol to replace the Luger P'08. Walther set to the task, modifying its PP pistol and going through various formats until the 9mm 'HP' (Heeres Pistole) was accepted, this then being designated the Pistole 38, or P38. Well made, attractively plated in matt black and very reliable, the P38 featured a safety

---

Walther P38
Date: 1938
Calibre: 9mm Parabellum
Operation: Short recoil
Weight: 0.8kg (1.9lb)
Length: 213mm (8.38in)
Muzzle velocity: 350mps (1150fps)
Feed/magazine capacity: 8-round detachable box
Range: 30m (98ft)

---

indicator pin that showed whether there was a cartridge in the chamber or not, and it also had an advanced double-action lock, which enabled the operator to fire the weapon from a hammer-down position with just a single pull on the trigger.

RIFLES

7.92mm (0.31in) Gewehr 98 (Gew 98) rifle

The standard Imperial German Army infantry rifle of World War I was the Gewehr 98. This 7.92mm (0.31in)-calibre single-shot, bolt-action rifle was built around Mauser's seminal bolt-action design, perfected in 1888. The interwar German Army covertly placed a very large number of these in storage in contravention of the Treaty of Versailles. After Hitler's 1934 repudiation of the Treaty, limited manufacture of the aged design recommenced and continued until 1941. During the massive expansion of the Germany Army that the Nazis implemented in the mid-1930s, these stocks of Gewehr 98s were used to equip many newly raised infantry units. As new small arms, like the Karabiner 98k, entered service, the Gewehr 98 rifles were relegated to equipping rear-area and security forces.

The rifle utilized a five-round magazine. The firer inserted the ammunition in clips into the magazine from above, which allowed half-empty magazines to be 'topped up'. With an overall length of 125cm (49in), the Gewehr 98 was a rather long and unwieldy weapon by the standards of the 20th century. It fired its round at a muzzle velocity of 640m/s (2100ft/s). A substantial number of vintage stockpiled Gewehr 98 rifles were issued to the German Volkssturm Home Guard Militia during the war's desperate last months, including several battalions that tried in vain to defend encircled Berlin in late April 1945.

Mauser Gewehr 98
Date: 1898
Calibre: 7.92mm (0.31in)
Operation: Bolt-action
Weight: 4.2kg (9lb)
Overall length: 1250mm (49in)
Barrel length: 740mm (29.1in)
Muzzle velocity: 640m/s (2100ft/s)
Feed/magazine: 5-round box magazine
Range: 500m (1640ft)

Gewehr 98
The Gewehr 98's distinctive small circular bolt dis-assembly disk is clearly visible in the centre of the wooden butt-stock, as is the sling base slot located just in front of this.

7.92mm (0.31in) German rifles compared						
	Description	Designed	In service	Overall length	Unloaded weight	Muzzle velocity
Gewehr 98	Rifle	1898	1898–1945	125cm (49in)	4.2kg (9lb)	640m/s (2100ft/s)
Karabiner 98a	Cavalry carbine	1898	1898–1945	110cm (43.3in)	3.63kg (8lb)	870m/s (2854ft/s)
Karabiner 98b	Rifle	1925	1925–1945	125cm (49in)	4.01kg (9lb)	785m/s (2575ft/s)
Karabiner 98k	Rifle	1925	1935–1945	111cm (43.7in)	3.9kg (8.5lb)	755m/s (2477ft/s)

7.92mm (0.31in) Karabiner 98b (Ka 98b) rifle

In the 1920s, the Gewehr 98 rifle underwent a series of minor design modifications that were the result of lessons the German Army had drawn from their experience of World War I. Slightly modified new ammunition was adopted; to optimize for this, the rifle received a new tangent rear sight. The resulting weapon was designated the 7.92mm (0.31in) Karabiner (Ka) 98b carbine. Although termed a carbine, the Ka 98b was actually the same length as the Gewehr rifle, at 125cm (49in). According to some authorities, it was so named because the Treaty of Versailles limited the number of rifles Germany could produce, but not the number of carbines.

In fact, the Ka 98b was an elongated modification of the original Karabiner 98a, the shortened cavalry carbine variant of the Gewehr 98 rifle. At 4.10kg (9lb), the Karabiner 98b was 0.2kg (0.4lb) lighter than the Gew 98; its other main differences were its altered sling swivels and a modified bolt-action design. Other than these details, externally the weapons appeared

Small arms

German troops crouch next to a flak half-track; most of the soldiers carry the Karabiner (Kar) 98 bolt-action rifle, the standard infantry personal weapon during the war; the soldier standing on the left instead carries an MP 38 sub-machine gun.

very similar to one another. The Germans commenced mass conversion and/or production of the Karabiner 98b in 1935 and continued throughout the 1930s. Subsequently, the weapon equipped many thousands of German rifle platoons throughout World War II. Increasingly, however, production shifted over to its successor, the Karabiner 98k rifle.

7.92mm (0.31in) Karabiner 98k (Ka 98k) rifle

After their 1935 repudiation of the Treaty of Versailles, the Germans introduced a new standard rifle, the 7.92mm (0.31in) Karabiner 98k – the 'k' standing for 'kurz' (short). This weapon had been in production under foreign licence in Czechoslovakia, Poland and Belgium since 1925. The Mauser-Werke factory at Oberndorf-am-Neckar commenced limited domestic production of the weapon in 1934. Subsequently, with other firms joining, the weapon was in full-scale mass-production from 1935 until the war's end in 1945. So many Ka 98s were eventually produced – some 14.1 million – that by 1942 it became the most common rifle in German service, equipping tens of thousands of infantry platoons. However, in 1944–45, repeated Allied aerial attacks severely disrupted

House clearing

Three German soldiers clear a house. The *infanteer* in the centre is armed with the ubiquitous Kar 98 rifle. In the foreground the soldier with the foliage-festooned helmet holds an MP 38 submachine gun.

Kar 98

In terms of external appearance the Kar 98 looked quite similar to the Gew 98, but there existed three main distinguishing features; the Kar 98 was shorter, it sported a slightly larger trigger guard, and its fore-stock was markedly shorter than its predecessor's fore-stock.

Mauser Karabiner 98k
Date: 1935
Calibre: 7.92mm (0.31in) Mauser M98
Operation: Bolt-action
Weight: 3.9kg (8.5lb)
Overall length: 1110mm (43.7in)
Barrel length: 600mm (23.62in)
Muzzle velocity: 745m/s (2444ft/s)
Feed/magazine: 5-round internal box magazine
Range: 500m (1640ft) + with iron sights

Street fighting
An early war two-man MG 34 team employ their weapon during urban fighting in an unusual doctrinal fashion; the firer holds the weapon from the kneeling position with the cylindrical ammunition box tucked under his arm.

Ka 98k production. Consequently, this weapon never entirely supplanted the Gewehr 98 or the Karabiner 98b rifles, and all three models equipped German field units until the war's end. Economies of scale under mass-production enabled the Germans to drive down the Ka 98's unit cost to just 70 Reichsmarks.

A development of both the Gewehr 98 and the Karabiner 98b, the 98k was shorter than both these models, being 110.7cm (43.6in) in overall length. Indeed, the rifle was very close to the overall length of the genuine carbine variant of the Gewehr 98, the Karabiner 98a. The weapon had the same five-round magazine as its predecessors. The weapon was renowned for its accuracy and had a normal effective of range of 500m (1640ft), which could be increased to 1000m (3280ft) with telescopic sights. In the hands of a skilled soldier, the rifle could deliver up to 15rpm. The weapon could be fitted with a variety of enhanced sighting devices, including the telescopic Zielernrohr 39 and 42 devices.

The Ka 98k could also mount several different grenade-launching devices, including the 1942-designed 3cm (1.18in) Schiessbecher discharger cup (of which 1.4 million were made). In 1944, the hard-pressed Germans introduced the 'War Model' design, a simplified Ka 98k variant that helped speed up production.

MACHINE GUNS

7.92mm (0.31in) Maschinengewehr 34 (MG 34) machine gun

In the 1920s, the German High Command concluded that having separate light (bipod-mounted) and heavy (tripod-mounted) machine guns was an inefficient use of their limited resources. The Germans therefore developed the then-revolutionary concept of a dual-purpose light/heavy machine gun. This led to the evolution of the 7.92mm (0.31in) Maschinengewehr 34 (MG 34), the world's first general-purpose dual-role machine gun.

Some 577,000 MG 34s were produced in 1934–45, although increasingly it was replaced in the field by its successor, the MG 42. In its light role, with bipod, the MG 34 weighed 11.5kg (25lb) and fired 50- or 250-round ammunition belts, or alternatively 75-round saddle drums. The weapon delivered an impressive theoretical cyclic fire rate of 800–1000rpm. The weapon had a tubular air-cooled barrel ventilated by round

MG 34
This cutaway artwork shows an MG 34 with a 50-round drum magazine fitted; the bullets in the drum can be seen, as can the barrel within the barrel sleeve; note how the bipod folds back along the bottom of the barrel.

Maschinengewehr 34 (MG 34)
Date: 1936
Calibre: 7.92mm (0.31in) Mauser
Operation: Recoil, air-cooled
Weight: 12.1kg (27lb)
Overall length: 1219mm (48in)
Barrel length: 627mm (24.75in)
Muzzle velocity: 762m/s (2500ft/s)
Feed/magazine: 50- or 75-round drum magazine or up to 250-round belt
Cyclic rate: 800–900rpm
Range: 2000m (6560ft)+

holes in the sleeve, rather than the heavy and cumbersome water jacket used on previous heavy machine guns. Air cooling, however, was less efficient than water cooling; thus the MG 34's barrel would soon overheat when delivering sustained fire. The Germans cleverly countered this by designing the weapon so that a well-trained two-man crew could change the barrel in seconds, usually having one or two spare barrels to hand.

The weapon sat on a bipod that could be fitted under the barrel at its end or further up nearer to the rear V-notched sight that projected up at the rear of the barrel sleeve. The weapon featured a wooden shoulder stock and a standard-shaped pistol grip. Obtaining a muzzle velocity of 755m/s (2477ft/s), the weapon could deliver its rounds out to an effective range of 2000m (6560ft). The barrel of the MG 34 (and its successor the MG 42) was also fitted with a muzzle flash suppressor that, when combined with flashless powder, made it difficult to detect the machine gun when it was fired.

The MG 34 became the standard firepower asset in a nine-man German infantry section, which existed to service and protect the weapon's fire effect. The primary MG 34 gunner was typically the section's most experienced

MG 34 AT DEMYANSK

The light-role MG 34 was also a powerful offensive weapon. In April 1942, for example, the SS *Totenkopf* Division was locked in bitter fighting around Demyansk on the Eastern Front. On 21 April, platoon commander Sergeant Fiedler found his command pinned down and facing destruction through determined enemy ripostes. In desperation, Fiedler and his crewmate charged the enemy positions, firing (in doctrinally bizarre fashion) their MG 34 from the hip on the move. In just three minutes, the pair, belching a storm of fire, assaulted three Soviet trenches; taken by surprise, 38 Red Army soldiers capitulated.

Light-role MG 34 and 42 compared			
	Barrel length	Weight incl. bipod	Muzzle velocity
MG 34	62.7cm (24.7in)	11.5kg (25lb)	755m/s (2477ft/s)
MG 42	53cm (20.9in)	11.6kg (25lb)	820m/s (2690ft/s)

soldier and one of the physically largest, as he had to lug the weapon around. His teammate fed the ammunition belt, periodically changed the barrel and cleared jammed rounds. In the defensive role, two other section soldiers brought up fresh ammunition for the gun while the other five deployed in foxholes to protect the machine-gun nest from enemy action. The impressive fire effect of the MG 34, particularly with sustained fire, goes a long way to explain the defensive resilience often seen by German units, particularly during the war's final two years. Just a few MG 34s, well sited to exploit the ground and with a copious supply of ammunition available, could bring to a halt the attacks of powerful Allied forces.

At the micro-tactical level, the potency of the MG 34 in part helps explain the combat effectiveness seen in many wartime German units.

Maschinengewehr 42 (MG 42) machine gun

While the MG 34's prowess was feared by Allied personnel, its performance

MG 34 in action

A three-man SS crew man an MG 34 machine gun in the heavy role; mounted on a large tripod. The Germans developed over a dozen different machine gun tripod designs.

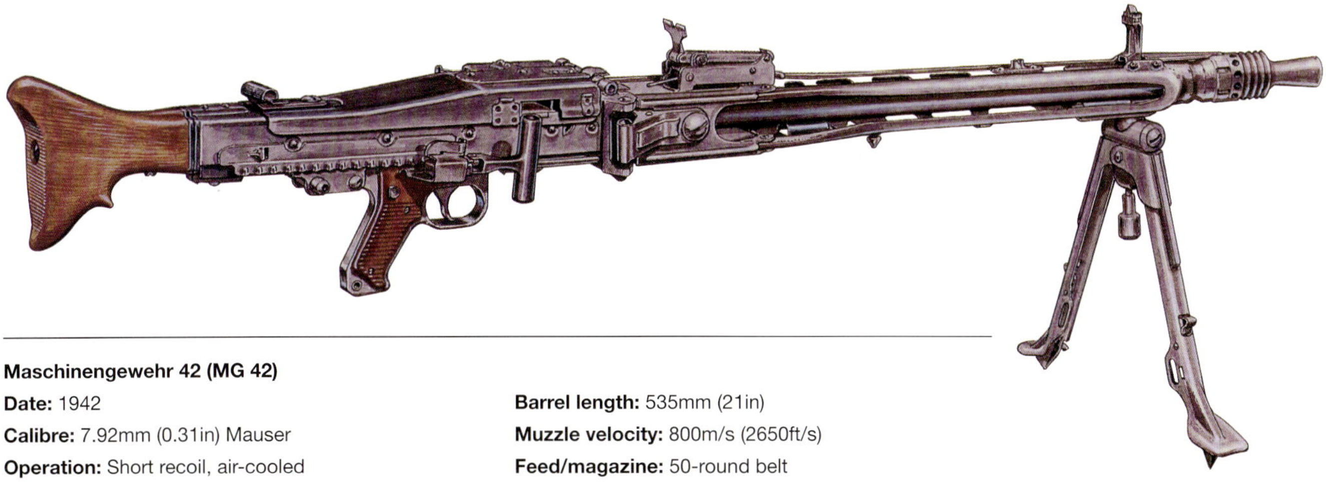

Maschinengewehr 42 (MG 42)
Date: 1942
Calibre: 7.92mm (0.31in) Mauser
Operation: Short recoil, air-cooled
Weight: 11.5kg (25lb)
Overall length: 1220mm (48in)
Barrel length: 535mm (21in)
Muzzle velocity: 800m/s (2650ft/s)
Feed/magazine: 50-round belt
Cyclic rate: 1200rpm
Range: 3000m (9842ft) +

MG 42
In this MG 42 cutaway the outer barrel cooling sleeve is omitted to reveal the barrel beneath it. This image illustrates how wooden components have been reduced to just butt-stock and pistol grip; everything else is easily-manufactured stamped metal components.

was eclipsed by its successor, the MG 42, now widely accepted as the most effective dual-purpose machine gun ever designed by that time. In 1940, Mauser began developing a more effective, but also easier to produce successor to the MG 34. Mauser applied their recent experience with cheap mass-production of the MP 38 and MP 40 sub-machine guns to create the MG 42, which was constructed using cheap and easy to produce die-cast and stamped components. Despite these basic building blocks, the MG 42 remained a superbly engineered weapon. It was differentiated from its predecessor by its rectangular barrel sleeve with six elongated oval ventilation holes along each side.

To maximize its sustained fire effect, the MG 42, like its predecessor, was optimized for swift barrel changes; thanks to a single spring-release mechanism, the crew could achieve this even more swiftly than on the MG 34. Swift barrel changing enabled the weapon to achieve what was for its day an astonishing theoretical cyclic sustained fire of rate of 1400–1500rpm.

The weapon had a hinged notch rear sight instead of the V-notch of its predecessor. The fire effect of just a couple of well-sited MG 42s with interlocking arcs was devastating for enemy assault forces attempting to close contact with these weapons' firing nests. The weapon fired the same 50- and 250-round ammunition belts as its predecessor. This meant that German infantry sections could operate either weapon with its allocated ammunition. Dubbed the dreaded 'Spandau' by the Allies, the weapon made a distinctive staccato 'brrpp' on firing that added to the already considerable morale-boosting effect of its fire.

The MG 42 could be used tactically in a heavy role mounted on a tripod, or permanently sited in a heavy role in a fixed installation, usually within

Defending the Reich
View of a late-war SS two-man MG 42 crew, one seemingly a young lad wearing just a cap, not a helmet. They man their weapon in a position surrounded by the bare twisted trunks of trees in front of an ironwork fence; note the gun's distinctive elongated oval barrel sleeve cooling holes.

an artillery battery or strongpoint. The Germans developed several tripods for the weapon, including the lighter Dreibein 34 mount and the Lafette 42. Several mountings enabled the weapon to operate in the anti-aircraft role, including the Zwillingslafette 36, which paired two MG 42s together. Several dozen large permanent pedestal mountings, often with shields to protect the crew, also existed for the heavy role MG 42 when sited within a fortified location. Most MG 42s in the heavy role incorporated a variety of sophisticated optical equipment, such as a dial sight.

The Germans manufactured 750,000 MG 42s in 1942–45. The weapon increasingly ousted the MG 34 from front-line service, although many units continued to field both weapons simultaneously. The MG 42's lethality helps to account for the formidable German defensive resilience displayed in the war's latter stages.

One such action occurred southwest of Caen during the British Epsom offensive, on 28 June 1944. Despite being assaulted by the two forward-deployed companies of a British infantry battalion, the four MG 42s of the 6th Company, 1st SS Panzergrenadier Regiment, unleashed sustained fire that pinned down the advancing enemy, subsequently prompting a withdrawal. The many thousands of MG 42s deployed across all fighting fronts meant that the Allies often had to fight protracted attritional all-arms warfare to gradually overwhelm the sheer volume of fire that these weapons delivered.

SUBMACHINE GUNS AND ASSAULT RIFLES

9mm (0.36in) Maschinenpistole 28 (MP 28) sub-machine gun

During World War I, a new small arm emerged, the sub-machine gun, in the form of the German Maschinenpistole (MP) 18 Bergmann. This was a light, short, easily portable, one-man weapon designed to deliver (not necessarily accurately) a high rate of automatic fire over a limited range. In the 1920s,

Warsaw ghetto
German troops search civilians in the aftermath of the AprilMay 1943 Warsaw Jewish ghetto uprising. The soldier in the foreground carries an MP 28 over his shoulder; note both the left-sided magazine and how at 82.8cm (32.8in) the weapon is relatively short.

the Germans developed an improved Bergmann variant, designated the MP 28. This weapon carried either a 20- or 32-round box magazine that the firer inserted into the weapon's horizontal magazine receiver from the left-hand side. The MP 28 was a well-constructed and rugged weapon, made more robustly than the later metal-stamped German sub-machine guns, the MP 38 and MP 40. Inevitably, this design made the weapon relatively heavy, at 5.2kg (11.5lb). This did not necessarily help with a weapon that was often fired in a 'spray and pray' fashion from the hip.

MP 28
The MP 28 derived from the MP 18 Bergmann, a World War I era sub-machine gun designed to deliver a rapid rate of fire over a short distance.

Maschinenpistole 28
Date: 1928
Calibre: 9mm (0.36in) Parabellum
Operation: Blow-back
Weight: 4.2kg (9.25lb)
Overall length: 815mm (32in)
Barrel length: 195mm (7.75in)
Muzzle velocity: 395m/s (1300ft/s)
Feed/magazine: 32-round detachable drum
Range: 70m (230ft)

The MP 28 was of the 9mm (0.36in) calibre that became standard for the genre. The weapon had a theoretical rate of fire of 350–450rpm, but this could not be sustained in actual situations. A relatively inaccurate weapon, the firer would usually deliver area-suppressive fire rather than aimed bursts, up to a range of 200m (656ft). The weapon incorporated both a wooden rifle-style stock and a perforated barrel sleeve to provide air cooling, since the barrel soon became hot with sustained fire.

9mm (0.36in) Maschinenpistole 38 (MP 38) sub-machine gun

This weapon was produced by the Erma-Werke in Erfurt, so the weapon became known to German service personnel as the 'Erma'; to the Allies, however, it was the 'Schmeisser'. The MP 38 proved to be one of the most effective sub-machine gun designs of World War II. It differed from previous German sub-machine guns in that it was purposely designed for cheap mass-production. Large parts of the weapon utilized die-cast or stamped metal components; these elements were produced by subcontractors and then dispatched to central locations for final assembly on an industrial-scale machine-tool production line. Only metal and plastics were used in its manufacture, with no wooden elements at all.

This 4.1kg (9lb) weapon utilized the standard 'blow-back' method of automatic fire common to this genre of weapons. The MP 38 also employed a 32-round rectangular magazine that was inserted vertically into the housing on the bottom of the barrel to create a foregrip for the firer.

The weapon fired solely in the automatic mode and in theory could

MP 38
The MP 38 had no safety catch to prevent the weapon being accidentally fired (for instance if the device was dropped); this omission was rectified in the modified subsequent MP 38/40.

Maschinenpistole 38 (MP 38)
Date: 1938
Calibre: 9mm (0.36in) Parabellum
Operation: Blow-back
Weight: 3.97kg (8.75lb)
Overall length: 833mm (32.8in) stock extended; 630mm (24.75in) stock folded
Barrel length: 251mm (9.9in)
Muzzle velocity: 395m/s (1300ft/s)
Feed/magazine: 32-round box magazine
Cyclic rate: 500rpm
Range: 60m (260ft)

achieve a maximum cyclic rate of fire of 450–500rpm. In actual tactical situations, German personnel tended to fire short bursts of 10 to 15 rounds, often in confined spaces such as house-to-house combat.

Uniquely for this genre of weapons up to that point in time, the lightweight skeleton butt-stock of the MP 38 folded up, enabling it to be stowed even in cramped conditions. A slightly modified variant, the MP 38/40, incorporated minor changes including an additional safety slot at the forward-firing position.

Large numbers of the MP 38 were issued to section and platoon leaders in infantry and mechanized divisions. Further large numbers were issued to personnel who worked within limited space, such as tank crews. Many German paratrooper formations also utilized the weapon in significant numbers, where its ease of stowage in confined spaces and portability were highly prized.

Maschinenpistole 40 (MP 40) sub-machine gun

The Germans commenced manufacture of the 9mm (0.36in) MP 40 sub-machine gun in the summer of 1940. This design was essentially an improved MP 38 that was optimized for cheap and swift large-scale manufacture; it was specifically designed to require the minimum number of different machine tools in its production.

The MP 40 weighed 4kg (9lb), fired its 9mm (0.36in) Parabellum rounds at a muzzle velocity of 390m/s (1280ft/s), and possessed an effective range of 200m (656ft); in this, its performance was identical to that of its predecessor, the MP 38.

The weapon was theoretically capable of cyclic automatic fire of up to 500rpm, though in real tactical situations (and with just a 32-round magazine) actual rates of fire were significantly lower; soldiers often 'spray-fired' the weapon from the hip or in the prone position.

Indeed, in combat, the MP 40 proved to be an effective weapon that was highly desired by troops. Like the MP 38, which it served alongside, the weapon was primarily issued to platoon and section leaders, tank crews and paratroopers. As the war progressed, its employment became much more

Winter campaign
The unexpected continuation of the Axis invasion of the Soviet Union into the winter of 1941–42 forced many expedients upon the Germany Army. Here a group of German infantry have donned some makeshift white winter camouflage over-garments. The trooper located second to the left carries an MP 40 slung over his shoulder.

MP 40

This artwork shows an early MP 40; its magazine receiver is not notched and the barrel tip is one solid unit not two. The weapon's folding butt-stock is bent back under the gun from the rear almost to the magazine/fore-handle.

Maschinenpistole 40 (MP 40)
Date: 1940
Calibre: 9mm (0.36in) Parabellum
Operation: Blow-back
Weight: 3.97kg (8.75lb)
Overall length: 832mm (32.75in) stock extended; 630mm (24.75in) stock folded
Barrel length: 248mm (9.75in)
Muzzle velocity: 400m/s (1312ft/s)
Feed/magazine: 32-round box magazine
Cyclic rate: 500–550rpm
Range: 80m (260ft)

widespread across the Wehrmacht, with the Waffen-SS in particular employing back-channels to acquire larger numbers of the weapon to part-equip some of its SS grenadier companies.

Despite a five-year production run, almost no modifications to the basic design emerged. In part this reflected the desire not to interfere with the rate of mass-production, but it is also testament to the effectiveness of the initial design.

The only modification of note was that the housing on the barrel for the ammunition magazine was changed in 1943 from a smooth finish to a ribbed one. Between 1940 and 1944, some 40 subcontractors and five main assembly plants produced 1,047,000 MP 40s.

7.92mm (0.31in) Machinenpistole 43, 43/1, 44 (MP 43, 43/1, 44); Sturmgewehr 44 (StG 44) assault rifles

During the latter half of World War II, the Germans pioneered the introduction of a hitherto unknown type of weapon: the automatic-fire assault rifle. In the 1920s, the Germans carefully analysed the lessons of World War I. These studies showed that most infantry fire-fights took place at ranges under 500m (1640ft), or about half the effective range of a rifle. What was needed, the Germans concluded, was an automatic self-loading rifle that could deliver rapid massed firepower across the 250–750m (820–2460ft) range. Throughout the interwar period, developmental work on such weapons was desultory. But under the crucible of total war, the Germans forged ahead in 1941–42, in part influenced by captured Soviet Tokarev automatic rifles, to produce the interim, but not entirely successful, Machinenkarabiner 42(H) machine-carbine.

Further developmental work continued; eventually, in late 1943, the Germans introduced into service the MP 43. This was later redesignated the MP 44 and then the Sturmgewehr 44 ('Assault Rifle' 44). The MP 43 was the first of the genre of small arms known as assault rifles, which are today used by every nation in the world; these weapons are high rate-of-fire self-loading magazine-fed automatic rifles.

The MP 43 weighed 5.22kg (11.5lb) and was 94cm (37in) in length. Utilizing a 30-round outwardly curling magazine (which acted as a foregrip), the self-loading MP 43 fired its 7.92mm (0.31in) x 33 'short' round with a muzzle velocity of 685m/s (2247ft/s). It could generate a theoretical cyclic rate of fire of up to 500rpm and a practical tactical rate of 120rpm. The MP 43/1 variant mounted a screw-on cap fitting to the barrel tip that fired grenades. Some late-production StG 44s manufactured in early 1945 also mounted the Zielgerat 1229 Vampir ('Vampire') infrared sight for night-time combat actions.

From late 1943 onwards, the MP 43/StG 44 began to reach frontline units, but it always remained a relatively uncommon weapon. At first the weapon equipped a solitary 'assault' platoon in a panzergrenadier company of just a select number of elite panzer divisions. Indeed, veterans would be cross-posted to this platoon to create a particularly well-trained sub-unit able to utilize the awesome firepower of the StG 44 to best effect. As 1944 unfolded, more StG 44s were allocated to select Waffen-SS mechanized and grenadier divisions.

MP 43
The world's first assault rifle, the MP 43. The small rod projecting forward from the front top of the rifle, just behind the fore-sight, is the gas block screw. The curved magazine fitted vertically into the receiver located in front of the trigger mechanism, and acted as a fore-handle.

Maschinenpistole 43 (MP 43)
Date: 1943
Calibre: 7.92mm (0.31in) Kurz
Operation: Gas-operated
Weight: 5.1kg (11lb)
Overall length: 940mm (37in)
Barrel length: 418mm (16.5in)
Muzzle velocity: 700m/s (2300ft/s)
Feed/magazine: 30-round detachable box
Cyclic rate: 550–600rpm
Range: 300m (984ft)

On retreat
A late-war German soldier, dressed in an unusual mixed combination of pieces of uniform, carries a StG 44 assault rifle slung over his shoulders. The weapon's slim, wedge-shaped safety catch is visible, located slightly behind and above the trigger mechanism.

By the end of the year, elite SS panzer divisions like the *Leibstandarte* and *Das Reich* each possessed either two platoons or an entire company within one or two chosen battalions. These units tended to be powerfully reinforced with mortars and employed as the divisional counter-attack reserve force.

In early 1945, for example, three companies of the 2nd Battalion, SS Panzergrenadier Regiment 26, part of the infamous 12th SS Panzer Division *Hitlerjugend*, were equipped with the StG 44. The Division was then deployed on the Eastern Front in the Hungarian border region. The battalion became the division's official 'Stoss' battalion. Unsurprisingly, when the *Hitlerjugend* went onto the offensive on 7 March to enlarge its bridgehead over the River Gran, this battalion successfully led the assault on the town of Puszta. The heavy firepower of the StG 44-equipped company played a key part in local successes.

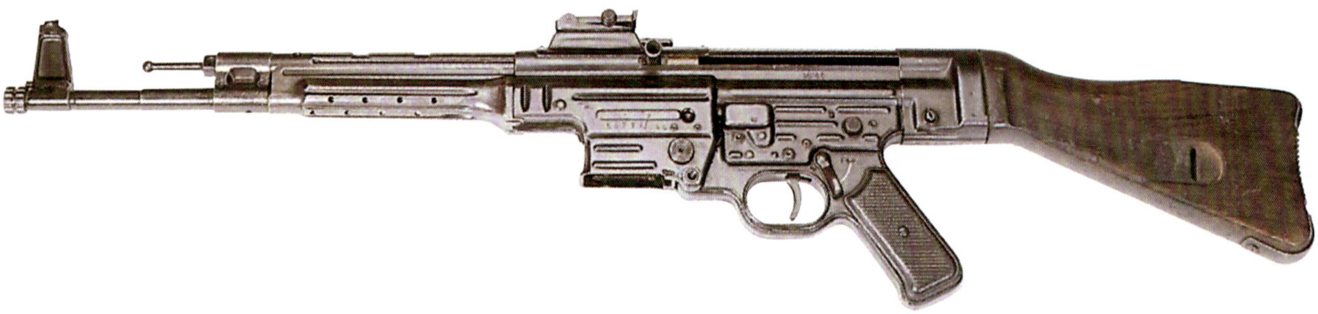

StG 44
Externally, the StG 44 was almost identical to the MP 43; it could, however, be distinguished from its predecessor by having a shorter barrel tip in front of the fore-sight as well as by sporting two adjoined barrel tip caps.

Sturmgewehr 44 (StG 44)
Date: 1944
Calibre: 7.92mm (0.31in) Kurz
Operation: Gas-operated
Weight: 5.1kg (11lb)
Overall length: 940mm (37in)
Barrel length: 418mm (16.5in)
Muzzle velocity: 700m/s (2300ft/s)
Feed/magazine: 30-round detachable box
Cyclic rate: 550–600rpm
Range: 400m (1312ft)

Fallschirmjägergewehr 42 (FG 42)
Date: 1942
Calibre: 7.92mm (0.31in) Mauser
Operation: Gas-operated
Weight: 4.53kg (10lb)
Overall length: 940mm (37in)
Barrel length: 502mm (19.76in)
Muzzle velocity: 761m/s (2500ft/s)
Feed/magazine: 20-round detachable box magazine
Range: 400m (1312ft)

FG 42
A single 7.92 x 57mm Mauser cartridge, fed from a horizontally-positioned left-hand 20-round rectangular magazine, sits in the rear of the chamber of this remarkable selective-fire assault weapon, the FG 42.

7.92mm (0.31in) Fallschirmgewehr 42 (FG 42) assault rifle

During the development of the MP 43, the Germans decided to design a similar weapon optimized to meet the specific needs of German paratroopers (Fallschirmjaeger), namely that it be a light, easily carried, readily stowed, high-firepower weapon. The ensuing design was designated the Fallschirmgewehr 42 (FG 42) assault rifle. Formed from lightweight materials, the rifle weighed just 4.5kg (10lb), more than 20 per cent less than the StG 44.

The weapon also had a lightweight fold-down bipod, enabling it to be used as a light machine gun. Utilizing side-mounted 20-round box magazines, the assault rifle could deliver 120rpm, but as a light machine gun it could achieve a theoretical cyclic rate of fire of 750–800rpm, with an effective range of 1200m (3940ft).

This was an expensive and complex weapon, and with German paratroop formations increasingly being used as ground-based infantry, it was produced only in small numbers – around 7100 examples. A good proportion of these were allocated to the elite 500th and 600th SS Commando-Paratrooper Battalions. The soldiers of the 500th SS Battalion used the FG 42 during their 25 May 1944 parachute drop on Communist guerrilla leader Josef Tito's mountain cave headquarters near Drvar. Although Tito managed to escape, the FG 42's firepower allowed the greatly outnumbered paratroopers to hold off partisan forces until a German ground relief column reached them.

ANTI-TANK WEAPONS AND GRENADES

Faustpatrone 'Panzerfaust' anti-tank launcher

By the mid-war years, the German Army had recognized that what they desperately needed was a light, man-portable, infantry anti-tank weapon that could be quickly produced in large numbers. Consequently, in 1942–43, the Germans developed a revolutionary new personal low-velocity discardable

Waffen-SS in Normandy
A Waffen-SS grenadier, possibly from the *Hitlerjugend* Division, waits in the relative safety found behind a classic Norman ditch and bank, with his Panzerfaust ready by his side, with which to engage enemy armour.

hollow-charge infantry anti-tank rocket-weapon. It was officially called the Faustpatrone ('Fist Cartridge'), but was dubbed by troops the Panzerfaust – literally 'Armoured Fist'. The discovery of hollow-charge warheads allowed the development of anti-tank warheads of much smaller size.

The first model, the Faustpatrone 30, entered service in late 1942. It was a very simple device that consisted of a hollow steel tube 36cm (14in) long, with a hollow-charge grenade fitted at one end and an explosive charge in the centre providing a counter-blast that projected to the rear to eliminate recoil, in a fashion that mimicked a recoilless gun. The very earliest version had a very powerful back-blast that required the weapon to be held at arm's length, which made it very difficult to aim; nearby troops had to be careful not to be caught in the blast. All other production models had an elongated tube that allowed the weapon to be tucked under the arm and fired, which greatly improved its accuracy.

Weighing only 5kg (11lb), the weapon's hollow-charge warhead could penetrate 140mm (5.5in) of armour plate sloped at 30 degrees, quite sufficient to disable both the T-34 and the KV-1. The grenade was stabilized in flight by fins that folded onto the shaft of the grenade as it was loaded and then sprang out once in flight. The biggest drawback of the weapon was its extremely limited range of only 30m (98ft); it took a brave and determined soldier to patiently wait out the approach of an enemy tank until it

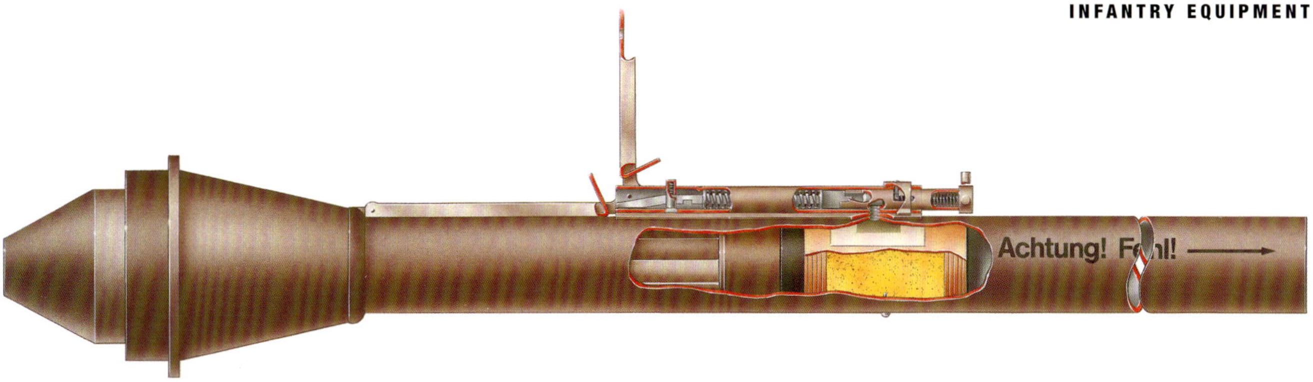

Panzerfaust
The crude sighting device on this Panzerfaust has been raised vertical, with the spring-loaded firing mechanism on top of the launcher tube located behind it; inside the launcher tube the stick-like bottom of the warhead can be seen, with its propellant charge.

Panzerfaust
Date: 1943
Calibre: 100mm (3.9in)
Operation: Recoilless gun
Weight: 1.475kg (3lb) total
Overall length: 1000mm (39.4in)
Muzzle velocity: 30m/s (98ft/s)
Feed/magazine: N/A
Range: 30m (98ft)

was within range. The earliest Panzerfaust weapons, therefore, had limited tactical utility. Nevertheless, the basic design proved sound and suggested possibilities for the future.

In 1943, the Germans developed the Faustpatrone 60, a weapon that could achieve a maximum range of 60m (197ft), double that of its predecessor. Subsequently, the Faustpatrone 100 device entered service in 1944; this had a double propellant charge that increased its effective range to 100m (328ft). During the last year of the war, the Panzerfaust was produced in huge quantities and widely issued to infantry tank destruction units. The Panzerfaust finally provided German grenadiers with the anti-tank capability for which they had long clamoured. During the last two years of the war, enemy armour thus became much more vulnerable to German infantry than they had hitherto whenever they operated without infantry support to suppress the threat posed by Panzerfaust-wielding enemy soldiers.

8.8cm (3.4in) Raketenpanzerbüchse 54 & 54/1 'Panzerschreck' anti-tank rocket-launcher

To augment the use of the Panzerfaust, the Germans also developed the hand-held Raketenpanzerbüshse (RPzB) 54 'Panzerschreck' ('Tank Terror') anti-tank rocket-launcher. The weapon was essentially an enhanced copy of the American 'Bazooka', which German troops had first captured examples of in Tunisia. The device consisted of an 88mm (3.4in) metal tube that fired a 3.3kg (7lb) fin-stabilized rocket-propelled hollow-charge anti-tank warhead at a muzzle velocity of 105m/s (346ft/s). The weapon could fire either the Panzerbüchsegranate 4322 or 4992 warheads to a maximum range of 150m (492ft). The easily manufactured weapon, constructed from simple lightweight materials, married a pistol grip and basic shoulder stock to the tube. A small protective shield was fitted in front of the trigger mechanism

Panzershreck
When fired the 'Panzerschreck' generated a lot of smoke from both its front and rear ends, which ran the risk of revealing the crew's position to the enemy; crews would thus prepare several firing positions and move between them after firing.

RPzB Panzerschreck
Date: 1943
Calibre: 88mm (3.4in) high-explosive (HE) and high-explosive anti-tank (HEAT) warheads
Operation: Solid rocket motor
Weight: 11kg (24lb) empty
Overall length: 1640mm (64.5in)
Muzzle velocity: 105m/s (346ft/s)
Feed/magazine: Breech-loader
Range: 150m (492ft)

and contained a small, round, transparent plate on the left-hand side to allow the operator to sight the weapon while the shield protected him; when the rocket was fired, its powerful flame back-blast posed a considerable risk to the crew. Indeed, many Panzerschreck operators also wore a gas mask and protective overalls just to be on the safe side.

A two-man crew serviced the weapon. The first crewman held and aimed the weapon, and the second primed the rocket and inserted it into the rear of the tube before attaching the initiator circuit wire; next, the first crewman fired the weapon. The Panzerschreck, however, was reusable and was, therefore, not discarded after firing.

Weighing 9.2kg (20lb) unloaded, the weapon proved more accurate than the Panzerfaust and had the advantage that it could be fired from a prone, rather than from a kneeling or standing, position. The RPzB 54/1 was a modified variant introduced in late 1944 that utilized a shorter tube; it could only fire the Panzerbüshsegranate 4992 rocket. Indeed, during the war's last six months, the Wehrmacht increasingly relied on small mobile tank destruction units, equipped with Panzerfaust and Panzerschreck to inflict heavy losses on massed Allied armour as it advanced inexorably deeper into the heart of the German Reich.

One such unit was the SS Tank Destruction Company Dora II, which played a key role in the German defensive stand made in front of Berlin on the Seelow Heights in 20–26 April 1945. The unit was equipped with StG 44 assault rifles, many Panzerschreck and Panzerfaust weapons, as well as anti-tank mines. In these desperate last stands, Dora II claimed to have destroyed 125 Soviet AFVs before its last surviving eleven soldiers were killed in a determinedly pressed Soviet attack on 28 April, southwest of Neu Zittau.

Panzerfaust and Panzerschreck compared						
	Anti-tank weapon type	Usability	In service	Muzzle velocity	Loaded weight	Maximum range
Panzerfaust 30	Hollow-charge launcher	One-shot discardable	1942–43	30m/s (98ft/s)	5.22kg (11lb)	640m/s (2100ft/s)
Panzerfaust 60	Hollow-charge launcher	One-shot discardable	1943–44	45m/s (148ft/s)	6.8kg (15lb)	870m/s (2854ft/s)
Panzerfaust 100	Hollow-charge launcher	One-shot discardable	1944–45	62m/s (203ft/s)	6.8kg (15lb)	785m/s (2575ft/s)
Panzerschreck	Rocket-launcher	Reusable	1943–45	105m/s (346ft/s)	11kg (25lb)	755m/s (2477ft/s)

GRENADES

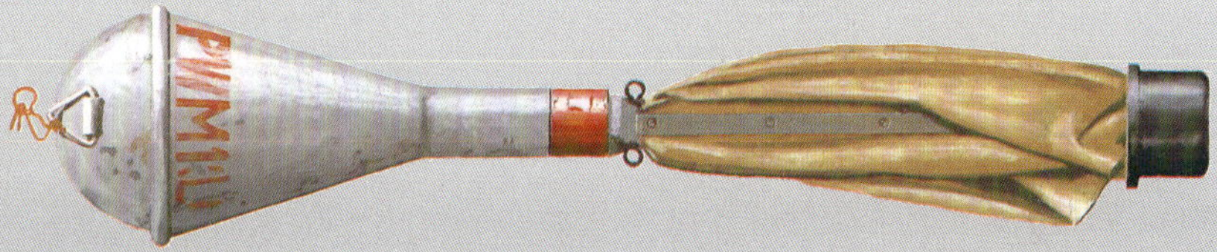

Panzerwurfmine
Date: 1944
Length: 533 (21in)
Weight: 1.35kg (3lb)
Explosive: 0.52kg (1lb) RDX/TNT
Armour penetration: 150mm (5.9in)

During World War II the German Army developed a range of explosive grenades designed to be thrown at close-range by troops at enemy soldiers. One common form of hand grenade employed was the egg-shaped 'Egg Grenade' (*Eihandgrenate*). The most common examples of this weapon were the Model 24 and Model 39 grenades. The latter remained in production from 1939 until 1945. This grenade was 7cm (3in) long and 5.1cm (2in) wide and could be fitted with 4, 4.5, 7.5 and 10 second delay fuses.

The Germans also produced a range of stick grenades (*Stielhandgrenate* – not illustrated here), which comprised a cylindrical warhead fitted to a short narrower cylindrical handle to enable the device to be thrown over greater distances; the Model 1924 and Model 1943 were the most commonly employed variants.

German troops also employed the *Panzwerwurfmine* (PWM), a hand-thrown shaped-charge anti-tank grenade. When a shaped-charge round hit an enemy vehicle, it emitted a high-velocity thin jet of metal that punched through the armour plating. For this to work effectively, the round should strike the enemy vehicle at 90 degrees. To help achieve this, the PWM had canvas sheaths fitted that helped stabilize its in-flight trajectory. In practice, this perfect 90-degree landing proved difficult to achieve in genuine tactical situations.

Model 39 Eihandgranate
Date: 1939
Weight: 230g (0.5lb)
Height: 76mm (3in)
Explosive: 112g (0.25lb) TNT
Filling weight: 112g (4oz)
Detonation mechanism: 4–5 seconds
Blast radius: 10m (33ft)

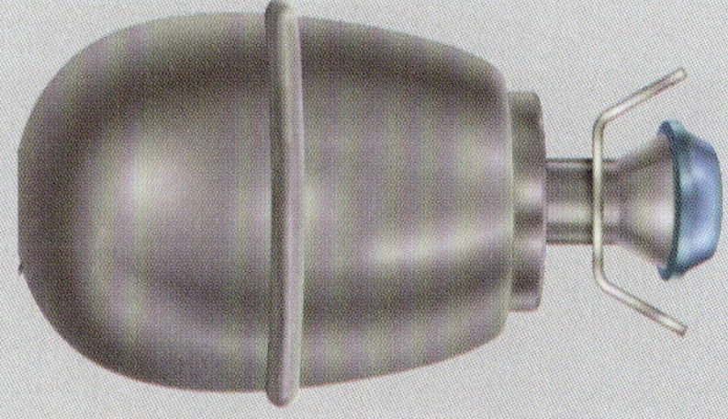

CHAPTER 4

Aircraft

The ability of the German ground forces to achieve tactical and operational success on the battlefield owed much to the efforts of Luftwaffe fighter squadrons to secure local aerial superiority. Luftwaffe domination of the skies played a crucial role in the early war victories in Poland, France and the invasion of the Soviet Union.

This was usually done through a 'counter-air campaign', in which German fighters contested 'dog-fights' with Allied fighters in the skies over the battlefields. Simultaneously, German fighters could be escorting bomber formations on their way to attack either Allied ground-force assets or air-base installations, such as runways and control posts.

German fighters also attempted to disrupt the increasingly numerous Allied strategic bomber attacks on installations within Germany or Nazi-occupied Europe.

FIGHTER AIRCRAFT

During the 1939–45 war, the Luftwaffe employed a large number of fighter models for such roles. Below we examine the four most important fighter aircraft types: the Messerschmitt Bf 109, the Me 110, the Focke Wulf Fw 190 and Me 410 Hornisse.

Opposite:
RAF recovery team
An RAF ground recovery team carefully inspect the wreckage of a downed Junkers Ju 88 A-series bomber; some 303 Ju 88s were lost during the fierce actions of the 1940 Battle of Britain.

Bf 109 G-6 fighter-bomber

This Bf 109 G-6/R-1 sports the R-1 factory modification kit, the ETX500-IXb bomb rack mounted under the fuselage between the landing gear. This device enabled it to carry a single SC250 250kg (550lb) bomb.

Messerschmitt Bf 109 fighter

First seeing combat in 1937 during the Spanish Civil War, the Messerschmitt Bf 109 was a low-winged, single-seat, single-engined, mono-tailed fighter that became the backbone of the German fighter force during World War II. In total, between 1936 and 1945, the Germans produced 33,984 Bf 109s, making it one of the most numerous fighter designs in aviation history. In addition to the Messerschmitt factory at Regensburg, six other facilities were involved in Bf 109 production. Although designed originally as a fighter-interceptor, later variants were optimized for the fighter-bomber, night-fighter, ground-attack and reconnaissance roles.

In 1933, the German Air Ministry published its requirements for the development of a single-seat short-range fighter. The first three slightly different Bf 109 prototypes underwent evaluation trials in 1935–36. These designs were compared with rival developments from Focke Wulf and Arado. The Bf 109 won the competition, and the Series A variants entered limited production in 1937.

Some 16 A-0 variants fought with the Condor Legion in Spain in 1937. The Series B was the first sub-type to enter general manufacture, with 341 B-1s produced in 1937–38. The B-1 was powered by a 492kW (661hp) Jumo 210D engine.

The Model C-1 entered service in early 1938 and was powered by an up-rated 514kW (690hp) Jumo 210G fuel-injection engine. This version featured strengthened wings and mounted four, rather than two, 7.92mm (0.31in) MG 17s. Just 58 examples of the four C series variants (C-1 to C-4) were manufactured in 1938.

The Series D Bf 109 became the standard production model prior to the start of World War II. The D-1 mounted two nose-mounted and two wing-mounted MG 17s and was powered by Jumo 210D engines. In total, six factories manufactured 647 Series D aircraft during 1938.

The next variant, the Model E series, entered production in late 1938. This model featured the more powerful 809kW (1085hp) Daimler-Benz DB601A engine and significantly redesigned and reinforced wings. The E-3 variant incorporated an enhanced armament arrangement of two MG 17s plus one MG FF cannon in each wing. The next sub-type, the E-4, of which 496 examples were built, incorporated the improved MG FF/M cannon. These early E series aircraft spearheaded the German efforts during the summer 1940 Battle of Britain, and suffered heavy losses in the process. The final E series model, the E-7, was designed to deliver greater operational range, thanks to modifications that enabled it to carry two external drop tanks; rather than the range of 660km (440 miles) obtained by the early Model Es, the E-7 could achieve a maximum range of 1325km (820 miles). In total, some 1196 Model

Messerschmitt Bf 109 E-1
Type: Single-seat fighter
Length: 9.02m (29ft 7in)
Wingspan: 9.92m (32ft 6in)
Height: 3.4m (11ft 2in)
Weight: 6600kg (14,551lb) maximum take-off
Powerplant: 1 x 809kW (1085hp) Daimler-Benz DB 601N inverted V-12 piston engine
Maximum speed: 570km/h (354mph)
Range: 720km (447 miles)
Service ceiling: 10,500m (34,450ft)
Armament: 4 x 7.92mm (0.31in) MGs; plus 4 x 50kg (110lb) bombs

Bf 109 E-1
Hauptmann Helmut Wick's Bf 109 E-1 served in I Gruppe (note sword insignia on nose) of Jagdgeschwader 2 ('Fighter Wing'), as indicated by the red 'R' in shield emblem, during the 1940 German invasion of the West.

Bf 109 G-6s patrol over Adriatic
Based on Crete, three Bf 109 G-6 aircraft bearing the markings of the 7th Squadron (white irregular emblem with apple inside located under cockpit) of the III Group (white shield with black cross on nose) of Jagdgeschwader 27 fly over the Adriatic, 1943.

Es were produced in 1938–40, including 438 E-7s.

The next major Bf 109 version to enter Luftwaffe service was the F series. These incorporated a redesigned fuselage that was aerodynamically more efficient, with numerous smoothed and rounded surfaces. The combination of better aerodynamics and improved, more fuel-efficient engines meant that with two drop tanks the F-1 could achieve a maximum range of 1700km (1060 miles). The F-1 also featured redesigned, quasi-elliptical wing tips. In terms of defensive armament, the Model F-1 mounted two synchronized 7.92mm (0.31in) MG 17s located above the engine and a nose-mounted 20mm (0.79in) MG FF/M Motorkanone (MK) auto-cannon that fired through the propeller hub.

The first F-1s produced got their baptism of fire during the tail end of the 1940 Battle of Britain. A total of 208 F-1s were built between August 1940 and February 1941. The F-2 sub-type mounted the 15mm (0.59in) MG 151 cannon instead of the MG FF/M. The ensuing F-3 to F-6 variants were all powered by the up-rated 993kW (1332hp) DB601-E engine, which increased the top speed they could obtain to 659km/h (410mph). The last three Series F sub-types mounted the 20mm (0.79in) MG 151/20 MK auto-cannon instead of the MG 151. Model F aircraft inflicted appalling losses on the often outdated Soviet airforce platforms they encountered during the summer 1941 Axis invasion of the Soviet Union.

The Bf 109 early G series (G1 to G4) utilized the basic Model F airframe with just some modifications. In production from 1942, the G-1 became the first model to incorporate a pressurized cockpit, and was followed by the G-3; these both served as high-altitude fighters. The G series versions were powered by the up-rated 993kW (1332hp) DB 605 V-12 engine. The G-2 and G-4 sub-types did not have pressurized cockpits and performed air superiority and fighter-bomber roles.

The 10 later G series sub-types (G5 to G-14) incorporated enhanced engines and armaments. The G-6, for example, utilized the 1084kW (1455hp) DB 605A engine, could obtain a top speed of 640km/h (398mph), and mounted three 20mm (0.79in) MG151/20 cannon and one 13mm (0.5in) MG 131. Increasingly from 1942 onwards, Bf 109 Models F-K deployed in the West were replaced by Focke Wulf 190s. Within the Reich, in the Mediterranean and on the Eastern Front, Bf 109s continued to dominate aerial combat.

The final Bf 109 production version was the K series, which entered manufacture in late 1944; around 3500 were produced before the war's end. These aircraft were powered by the enhanced 1323kW (1775hp) DB 605D engine. These aircraft had further aerodynamic improvements that kept the design's rate of climb superior

Bf 109 annual production statistics

Year	Number produced
1935–38	1860
1939	1540
1940	1868
1941	2628
1942	2658
1943	6418
1944	14,152
1945	2800
Total	33,984

Bf 109 F-4

This Bf 109F-4 was flown by Hauptmann Hans-Joachim Marseille, who claimed 158 official victories before his death west of El Alamein on 30 September 1942. On one occasion, Marseille scored 14 victories in a single day. No other pilot claimed as many Western Allied aircraft as Marseille.

to that of the latest Allied Mustang, Spitfire and Tempest models. Both the widespread use of the Bf 109 and its effectiveness is attested by the fact that the 105 leading Bf 109 pilots who claimed more than 100 'kills' each alone accounted for some 15,000 enemy aircraft destroyed.

Messerschmitt Me 110 heavy fighter

Designed in 1934–36, the Me 110 was a two-crew, two-engined monoplane heavy long-range escort fighter. Powered by a Daimler-Benz DB 600 engine, the first prototype flew in May 1936 and obtained a maximum speed of 509km/h (316mph). The first production variant, the Series B, was intended to mount the DB 600 engine, but the technical problems experienced with it led to the substitution of the less powerful Juno 210B engine, which reduced the Model B's speed. The B-1 mounted by way of defensive armament two 20mm (0.79in) MG FF cannon, four MG 17s and a single rear-firing MG 15.

The next production series was the Model C, of which seven sub-types (C-1 to C-7) were manufactured. The C-1 featured the improved and more reliable 824kW (1105hp) DB 601A engine, as well as the same armament as the B-1. Entering service in summer 1939, the B-1 got its combat baptism of fire during the Polish campaign. The C-3 featured the improved MG FF/M cannon, and 195 of these planes participated during the 1939 Polish campaign.

During the May–June 1940 invasion of the West, some 60 Models B–C were lost. The C-4, which experienced its first combat missions during the 1940 Battle of Britain, sported armoured plates in the crew cockpit. The Model D was a long-range version of the C series that was employed during the spring 1940 Norwegian campaign. With

Eastern Front patrol
This Me 110 C-1 aircraft '3U+KK' of the 9th Squadron, Zerstörergeschwader 26 leads a stick of at least two other Me 110s flying over Wilno, Poland, during the 1941 Axis invasion of the Soviet Union.

Messerschmitt Bf 110 G-4b/R3

Type: Twin-seat fighter

Length: 12.65m (41ft 6in)

Wingspan: 16.27m (50ft 3in)

Height: 3.5m (11ft 6in)

Weight: 6750kg (14,881lb) maximum take-off

Powerplant: 2 x 1115kW (1496hp) Daimler-Benz DB 605 inverted V-12 piston engine

Maximum speed: 562km/h (349mph)

Range: 775km (482 miles)

Service ceiling: 10,900m (35,750ft)

Armament: 2 x 20mm (0.79in) cannon; 4 x 7.92mm (0.31in) MGs

Bf 110 G-4b/R3

This Me 110 G-4/ R-3 night fighting aircraft sports in its nose the distinctive antenna for its FuG 202/220 Leichtenstein radar; this aircraft served with the 5th Squadron of Nachtjagdgeschwader 5 (formation 'C9') in the defence of the Reich during 1944.

Bf 110 C-4 and G-2		
Attribute	**C-4**	**G-2**
Loaded weight	6700kg (14,771lb)	7790kg (17,158lb)
Maximum speed	560km/h (348mph)	595km/h (370mph)
Engine power	809kW (1085hp) x2	1085kW (1455hp) x2
Power to mass	0.241kW/kg (0.147hp/lb)	0.279kW/kg (0.17hp/lb)

237 Me 110s A–Cs initially committed to the Battle of Britain, 223 were lost. All three of these 1940 campaigns highlighted the Me 110's grave limitations when dog-fighting with the latest Allied fighters.

The next development of the design was the E series, which entered production in spring 1941. This was powered by the more potent DB 601N engine, which enabled this model to carry up to 1400kg (2656lb) of internally and externally carried bombs. After just four months in production, it was replaced by the Series F variants. Powered by the improved 1006kW (1350hp) DB 601F engine, the F-1 incorporated reinforced cockpit glass. F series production was supposed to be phased out in favour of the Me 210, but when this and the Me 110 G both experienced technical difficulties, Series F manufacture was continued. The F-1 was a fighter-bomber variant, the F-2 was a heavily armed anti-bomber platform, and the F-4 was a night-fighter.

The Me 110 G only went into production in late 1942 after the failure of the Me 210 design project. The Series G aircraft differed from their predecessors by being powered by the 1115kW (1496hp) DB 605 engine, which gave it a top combat speed of 562km/h (349mph). The G-2 was a heavily armed Zerstörer ('destroyer') variant, often employed in the ground-attack role against enemy AFVs, particularly on the Eastern Front in 1943. This sub-type sported two nose-mounted 20mm (0.79in) MG151/20 cannon and a MG81Z twin machine gun.

The most effective of the G series aircraft was the G-4. Designed as a dedicated night-fighter, the G-4 sported a FuG 202 or 222 radar and could be fitted in the factory or in the field with a wide range of conversion kits, including armament, bombload or radar upgrades. In summer 1943, Model G-equipped squadrons were redeployed as daylight anti-bomber assets against American strategic bombers, but suffered extremely heavy losses in the process. The G-4, however, subsequently proved effective as a night fighter engaging RAF strategic bombers. Production ended in August 1944 after 6170 examples had been manufactured.

Focke Wulf Fw 190 fighter

Designed in 1937–39, the first five Focke Wulf Fw 190 prototypes were subjected to extensive testing in 1939–40; the fifth one was powered by the potent two-row 1148kW (1540hp) BMW 801 14-cylinder engine. The first production model, the A-1, entered manufacture in June 1941 and was powered by the up-rated 1147kW (1539hp) BMW 801-C engine.

The A series aircraft packed more armament than the Bf 109 Model Fs then in service. The A-1 mounted two synchronized 7.92mm (0.31in) MG 17s (one in each wing root and two in the forward fuselage), together with two free-firing 20mm (0.79in) MG FF/M cannon (with one in each wing root). During the second half of 1941, Model A aircraft operated from French airbases and proved themselves more than a match for the latest Spitfire variant, the Mk.V. Indeed, a good number of German pilots believed that the Fw 190 was marginally superior to the Bf 109.

The Fw 190 A-3 sub-type was another up-gunned variant, which had two of its predecessor's MG 17s replaced with the more potent 20mm (0.79in) MG151/20E cannon. Several hundred A-1, A-2 and A-3 variants were committed to the Eastern Front in late 1942. Another subsequent sub-type, the A-5, saw the engine repositioned slightly further forward, increasing the maximum bomb payload that could be carried. Some 260 Fw 190s, including many A-1 to A-5 platforms, supported Operation 'Citadel', the July 1943 German armoured offensive at Kursk. Entering service in November 1943, the A-6 was another up-gunned sub-type intended for destruction missions against Allied heavy bombers; it sported six 20mm (0.79in) MG 151/20 cannon.

Fw 190-0s, summer 1941
Three Fw 190 pre-production aircraft of Eprobungsstaffel 190 ('Testing Squadron' 190) at le Bourget airfield, Paris, in Nazi- occupied France during summer 1941.

Entering mass production in February 1944, the A-8 variant featured the further up-rated 1456kW (1953hp) BMW 801-D-2 engine and heavy firepower capabilities similar to that of the A-6. Around 500 Model Fw 190 As spearheaded the German counter-air campaign mounted during the summer 1944 Normandy campaign; nearly 300 Fw 190s were lost in this intensive struggle. Some 380 Fw 190 As and 240 Ds also participated in Operation 'Bodenplatte', the massed German attack on Allied airfields mounted on 1 January 1945; 62 A-8s and 50 D-9s were lost during this operation. In total, between 1941 and the war's end in May 1945, German factories manufactured 13,291 Fw 190 Series A aircraft.

The chief concern with the A series aircraft was that their performance tailed off at altitudes above 6000m

Fw 190 A-1

This Fw 190 A-1 fighter is flown by an experienced pilot, indicated by the 19 vertical 'kill' markings on the tail; note the distinctive inward-sloping undercarriage arrangement, so typical of this design.

Focke Wulf Fw 190 A-1
Type: Single-seat fighter
Length: 8.84m (29ft)
Wingspan: 10.5m (34ft 5in)
Height: 3.96m (13ft)
Weight: 4900kg (10,803lb) maximum take-off
Powerplant: 1 x 1147kW (1539hp) BMW 801C radial piston engine
Maximum speed: 624km/h (388mph)
Range: 900km (560 miles)
Service ceiling: 11,400m (37,402ft)
Armament: 4 x 7.92mm (0.31in) machine guns; 2 x 20mm (0.79in) cannon

Fw 190Fs, Eastern Front
A view of two early F-series Fw 190 ground-attack variant fighters in flight, with visible the external bomb racks that the F-series incorporated.

(20,000ft). To rectify this, Focke Wulf developed the D series in 1943–44, which featured pressurized cockpits and a turbo-charged DB 603 engine. There were some 13 sub-types of this series (D-1 through to D-13); the D-1 entered service in September 1944. Some of these D type platforms incorporated further improved engines, such as the type D-11 with the improved Jumo 213F engine. Other sub-models had armament improvements; the D-11, for example, sported two 20mm (0.79in) and two 30mm (1.18in) cannon, while the D-13 sported a third 30mm (1.18in) cannon.

The next general-production series was the Model F, which was based on the A series, but with minor modifications for the ground-attack role. With some 432 examples manufactured, the F-3, for example, mounted one fuselage external bomb rack and two external wing-mounted bomb racks. The F-9, an adaptation of the A-9, sported a redesigned 'bulged' canopy and four wing-mounted external racks. Around 350 F-9s were completed in the

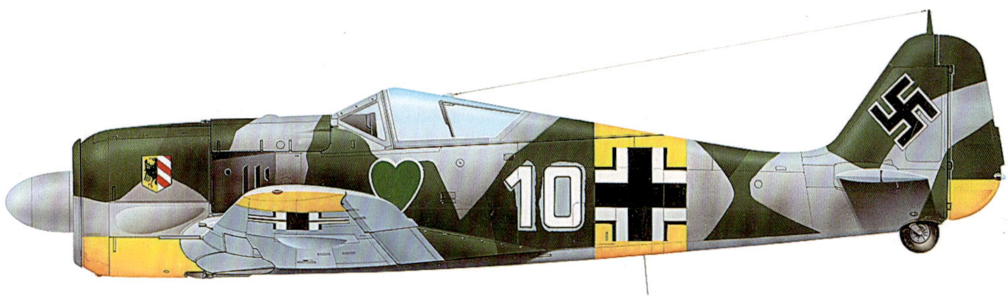

Fw 190 A-4
An Fw 190A-4 of Leutnant Walter Nowotny. Probably the most famous Fw 190 pilot, Nowotny was the first past the 250-kill mark, achieving 258 by the time of his death. Most of these were scored flying the 'Butcher Bird' with JG 54.

Fw 190 F-8, Schlachtgeschwader 2, Hungary 1942

This Fw 190 F-8 ground-attack variant could carry a range of bombs including one 500kg SC500K or eight 50kg SC50J devices. Note the white disruptive camouflage on the propeller nose cone.

Focke Wulf Fw 190 F-8

Type: Single-seat fighter

Length: 9m (29ft 5in)

Wingspan: 10.5m (34ft 5in)

Height: 3.95m (12ft 8in)

Weight: 4900kg (10,803lb) maximum take-off

Powerplant: 1 x 1268kW (1700hp) BMW 801D 18-cylinder radial piston engine

Maximum speed: 653km/h (408mph)

Range: 900km (560 miles)

Service ceiling: 11,400m (37,402ft)

Armament: 2 x 20mm (0.79in) MG 151/20 cannon in the wing roots; 2 x 13mm (0.51in) MG 131 MGs above the engine; 2 x 1800kg (3968lb) bombs

first four months of 1945. The ensuing series, the Model G, was optimized from scratch for the long-range fighter-bomber ground-attack role. Most of this type carried four external racks that could carry either additional fuel tanks or bombs. In total, 1300 Fw 190 Model Gs of all types were manufactured.

The Focke Wulf Ta 152 H was a late-war development of the Fw 190 intended as a high-altitude fighter that could take on the American B-29 Superfortress heavy bomber. This longer-nosed pressurized cockpit variant entered service in January 1945. The design was heavily armed, with one propeller hub-mounted 30mm (1.18in) MK 108 MK cannon, and two synchronized 20mm (0.79in) MG 151/20 cannon in the wing roots.

The Ta 152 had one of the most impressive performances of any piston-engined aircraft of the 1939–45 war. Its top speed was 755km/h (469mph) at altitude or 560km/h (350mph) at sea level. Only small numbers of Ta 152 Hs – probably 70 – reached operational units. Indeed, the number of operational examples in service peaked in March 1945 at a paltry 16. These aircraft shot down seven Allied aircraft for the loss of four Ta 152s.

Messerschmitt Me 410 Hornisse ('Hornet') heavy fighter

Known as the 'Hornet' (Hornisse), the two-crew Messerschmitt Me 410 heavy fighter was a development of the unsatisfactorily unstable Me 210, which never entered general production.

A two-engined monoplane, the Me 410 had large streamlined engine nacelles situated well forward of the wing leading edge, a large glazed crew canopy positioned high upon the front fuselage, and a large single tail piece. In comparison with the Me 210, the Me 410 featured a lengthened fuselage and redesigned automatic wing leading edge slats. The aeroplane was powered by the up-rated 1290kW (1730hp) Daimler-Benz DB 603A engine mounted in the fuselage nose. This propulsion system enabled the Me 410 to obtain a top speed of 626km/h (388 mph) and a maximum service range of 2300km (1429 miles). In total, German factories produced some 1160 Me 410s of all types during 1942–44.

The first production series, the Model A, mounted as defensive armament two nose-mounted 7.92mm (0.31in) MG 17 machine guns and two nose-mounted 20mm (0.79in) MG 151/20 cannons. In addition, many examples also featured two remotely controlled rear-facing 13mm (0.5in) MG 151s in fuselage side blister mounts. The A-1 was a light fighter-bomber variant that could carry a bomb payload of 1000kg (2204lb). This model could obtain an impressive maximum rate of climb of 9.3m/s (30.6ft/s). The A-3 sub-type was a dedicated reconnaissance variant with several cameras fitted. The second production series, the Model B, mounted as standard two potent

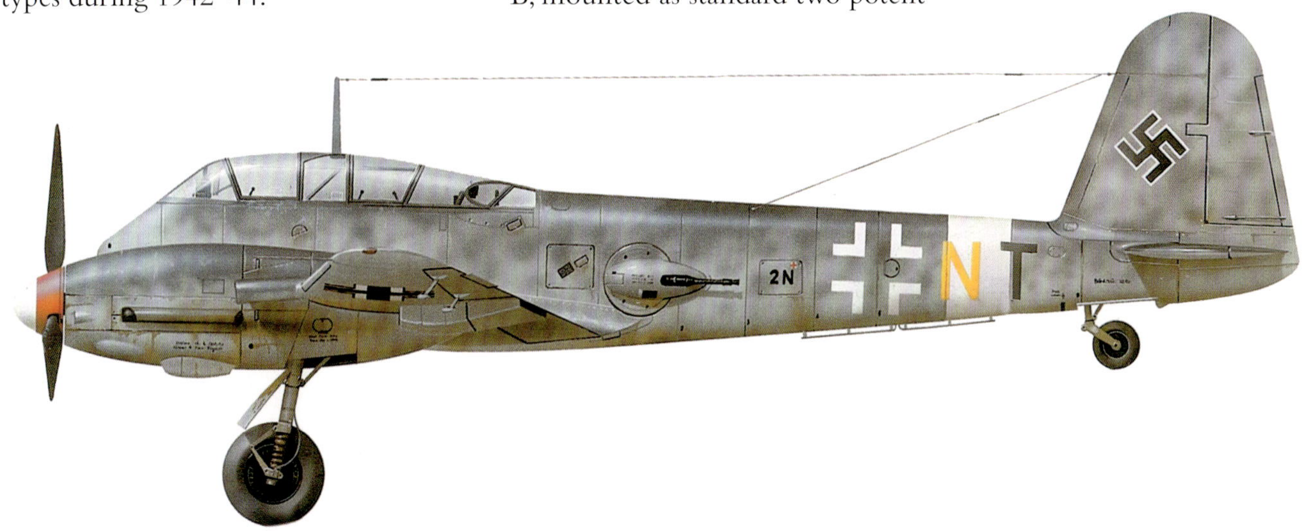

Messerschmitt Me 410 A-2/U2
Type: Single-seat fighter
Length: 12.48m (40ft 11in)
Wingspan: 16.35m (53ft 7in)
Height: 4.28m (14ft)
Weight: 9651kg (21,276lb) maximum take-off
Powerplant: 2 x 1380kW (1850hp) Daimler-Benz DB 603A inverted V-12 piston engine
Maximum speed: 624km/h (388mph)
Range: 1690km (1050 miles)
Service ceiling: 10,000m (32,800ft)
Armament: 2 x 20mm (0.79in) MG 151 cannon; 2 x 20mm (0.79in) MG 151 cannon in ventral tray; 2 x 7.92mm (0.31in) MGs in nose; 2 x 13mm (0.5in) MG 151 MGs mounted in rearward facing barbettes

Me 410 A-2/U2
The left-side rearward-facing MG 151 machine gun can be clearly seen in its blister mounting on the fuselage side behind and above the wings on this Me 410 A-2/U-2, fitted with the conversion kit U-2, namely two ventral MG 151/20 cannon.

Test flight
This view of a Me 410 heavy fighter ably illustrates three of the design's key features; its large nose, the extensive glassed cockpit area for the two-man crew, and its sporting of many rounded aerodynamically-optimised surfaces.

13mm (0.5in) MG 151 machine guns. The B-1 sub-type was the standard heavy fighter variant. Some B-1s were upgraded to B-1/U-2 or U-4 status by incorporating two 20mm (0.79in) MG 151/120 cannons or two 30mm (1.18in) MK 103 cannon in the under-fuselage ordnance bay.

The Hornet was employed in several tactical roles. In 1943–44, a key Me 410 role was mounting anti-bomber missions. For these sorties, a number of Me 410 As and Bs were converted in the field (receiving the suffix U-1 or U-2) to mount two 20mm (0.79in) or 30mm (1.18in) cannon in the under-fuselage ordnance bay. The Me 410 proved fairly successful in 1943 as a 'killer' of Allied bombers. However, by 1944, the number of long-range Allied fighters (such as the Mustang) protecting the bombers increased sharply. The Hornet could not compete with these fighters in dog-fights, and this led to significant losses among Me 410-equipped units. Increasingly in 1944, the remaining Me 410s were withdrawn from anti-bomber duties and instead used for night reconnaissance sorties.

GROUND-ATTACK AIRCRAFT

One of the roles that air power provides to support ground combat is 'close air support' – the provision of fire support to the contact battle, otherwise known as 'ground attack'. During the early phase of the war, ground attack was delivered mainly by the Ju 87 Stuka dive-bomber, although other fast light bombers such as the Do 17 performed these missions as well.

With the emergence of the conflict on the Eastern Front during mid-1941 onwards, however, the nature of ground-attack missions evolved to focus increasingly on destroying enemy armoured fighting vehicles (AFVs).

Henschel Hs 123

A single-seat biplane, the Henschel Hs 123 was employed in the early part of the 1939–45 war as a dive-bomber and close-support ground-attack aircraft. The platform was designed in 1933–35 to meet the Air Ministry's requirement for a single-seat biplane dive-bomber. During 1935–36, three prototypes were subjected to extensive testing. These all-metal prototypes featured 'sesquiplane' wings, with the lower wings being smaller than the upper ones. In these trials, the HS 123 proved to have excellent manoeuvrability that included being able to swiftly pull out of a near-vertical dive.

The Hs 123 A-0 pre-production variant entered limited manufacture in 1936, followed by the A-1 production sub-type. Powered by the 596kW (800hp) BMW 132 Dc9 engine, these A-1s sported two 7.92mm (0.31in) MG 17 machine guns mounted in the nose and synchronized to fire through the propeller. This model carried one 250kg (550lb) bomb under the fuselage centre-line and also sported two lower wing-mounted bomb racks that each carried two 50kg (110lb) bombs. Five Hs 123 A-1s were deployed to the Condor Legion, where they gained combat experience fighting in the Spanish Civil War. The Hs 123's relatively slow speed of 341km/h (211mph) meant that the

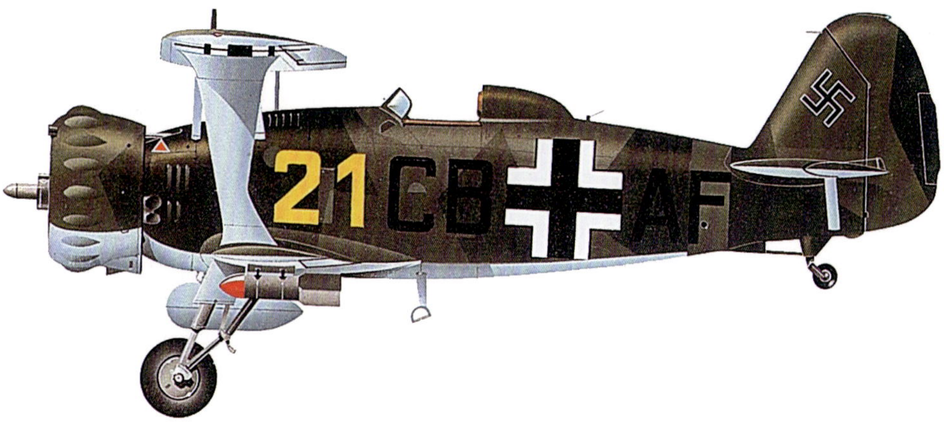

Henschel Hs 123 A-1
Type: Single-seat bi-plane
Length: 8.33m (27ft 4in)
Wingspan: 10.5m (34ft 5in)
Height: 3.22m (10ft 7in)
Weight: 2217kg (4888lb) maximum take-off
Powerplant: 1 x 596kW (800hp) BMW 132 Dc9 nine-cylinder radial engine
Maximum speed: 341km/h (212mph)
Range: 860km (534 miles)
Service ceiling: 9000m (29,520ft)
Armament: 2 x 7.92mm (0.31in) MGs; rack of 4 x 50kg (110lb) bombs

Hs 123 A-1
In addition to their ground attack role, some of the small numbers of Hs 123 A-1 aircraft remaining in service were used for pilot training duties. In this role, the planes often had their entire fuselage nose cowling painted yellow.

Hs 123 A-1, Moscow, 1941/42
This all-white Hs 123A-1 fought on the Central Moscow front over the winter of 1941–42. For winter operations, a white soluble distemper was applied over the permanent camouflage. The yellow band is a theatre marking.

aircraft proved accurate in delivering its bombs. The design also showed that it could take a lot of punishment from enemy guns and keep flying.

The first Ju 87 Stuka dive-bombers entered service in 1937; as these were superior in performance, the HS 123 A-1s in service were gradually reallocated to second-line duties. By the start of the September 1939 invasion of Poland, however, the Luftwaffe still had 39 Hs 123s in service. Although by no means reflecting state-of-the-art aerial technology, the Hs 123 proved to be effective in the ground-attack role in Poland. These aircraft were accurate in delivering their bombs and again demonstrated their ability to take a lot of damage from enemy flak but keep flying. In addition, ground crews found them easy to maintain and mechanically reliable.

The small number of Hs 123s still in service continued to deliver effective performances during the 1940 Western campaign and the spring 1941 Balkans campaign. On the Eastern Front in 1941–42, a single German ground-attack wing operating with 22 HS 123s and 38 Messerschmitt Bf 109 Model Es mounted hundreds of sorties against Soviet positions. Astonishingly, the dwindling number of antiquated Hs 123s continued to carry out ground-attack missions in the East until 1944, when they were withdrawn for glider-towing duties.

Henschel Hs 129

In 1937, the Air Ministry put forward a requirement for the development of a dedicated twin-engined, well-protected, low-level monoplane ground-attack aircraft. Three Hs 129 prototypes underwent evaluation trials in 1939–40. The design comprised a single 'bath-tub' sheet of metal that formed the aircraft's nose and cockpit area; the latter's glazed windows were formed of thick hardened glass. The fuselage was of an unusual triangular profile. Testing showed that although the general concept was sound, the plane was difficult to fly.

The first 16 modified A-1 production aircraft entered manufacture in summer 1940, powered by the 342kW (459hp) Argus As 410 inverted V-12 engine. These aircraft sported two 20mm (0.79in) MG151/20 cannon and two 7.92mm (0.31in) MG 17 machine guns; later aircraft were modified in the field

Henschel Hs 129 B-2/R2
Type: Single-seat ground attack
Length: 9.75m (31ft 9in)
Wingspan: 14.2m (46ft 5in)
Height: 3.25m (10ft 6in)
Weight: 5250kg (11,574lb) maximum take-off
Powerplant: 2 x 522kW (700hp) Gnome-Rhone 14-cylinder radial engine
Maximum speed: 407km/h (253mph)
Range: 688km (427 miles)
Service ceiling: 9000m (29,520ft)
Armament: 2 x 20mm (0.79in) cannon; 2 x 7.92mm (0.31in) MGs; heavy cannon or multiple MG pod, or up to 250kg (550lb) bombs under fuselage

Hs 129 B-2/R2
Two-view of a Henschel Hs 129 B-2/R-2 ground-attack aircraft, fitted with the Rüstatz-2 field modification kit with two 20mm (0.79in) MK 103 cannon; it has yellow finish applied to its nose, wing tips and at the vertical theatre-recognition stripe position in the rear fuselage.

to mount the 30mm (1.18in) MK 101 cannon instead of the 20mm (0.79in) MG151/20 guns. The A-1 design incorporated four bomb racks mounted under the fuselage's central line that could house four 50kg (110lb) bombs or 96 2kg (4lb) bomblets. However, these 16 were never completed as A-1s, as a more powerful engine had become available. These incomplete aircraft were converted to the B-0 and B-1 designs.

The B-1 variant began to enter service late in 1941, with the first examples being converted A-1s. These B-1s incorporated a more extensive glazed area in the cockpit, which increased pilot situational awareness. The aircraft were powered by the up-rated 522kW (700hp) Gnome-Rhône 14M radial engine. In early 1942, these Henschel Hs 129B-1s joined the Hs 123 and Bf 109E aircraft already serving with the 1st Group of Schlachtgeschwader 1 (Ground-attack Wing 1) on the Eastern Front. In May 1942, the first slightly modified B-2s joined this same wing. These aircraft could be modified in the field to mount the faster-firing 30mm (1.18in) MK 103 cannon.

As 1942 unfolded, it became clear that the MK 101 and 103 were proving less effective against the up-armoured Soviet tanks then reaching the front line. Consequently, some B-2s were converted in the field to mount the 3.7cm (1.46in) BK 3.7 auto-cannon instead. Other B-2s were built with 'tropicalized' modifications to optimize them for service in North Africa.

Henschel developed the B-3 sub-type to solve the challenge of destroying the latest heavily armoured Soviet tanks. This mounted the large and potent 7.5cm (2.96in) Bordkanone 7.5 auto-cannon in an under-fuselage weapons pod; the gun was fed by a 12-round rotary magazine. The BK 7.5 was the largest-calibre gun to be mounted on a Luftwaffe platform. While an undoubted 'tank-killing' weapon, incorporation of the heavy

Hs 129 B-3 fitted with Panzerabwehrkanon 40

This Henschel Hs 129 B-3 mounts under its front fuselage the powerful and bulky 7.5cm (2.95in) Bordkanone 7.5 cannon, the largest anti-tank weapon fitted to a German aircraft; the cannon's sheer weight made the B-3 challenging to fly.

Ju 87 G-1
This Stuka was flown by Major Theo Nordmann over Soviet Russia in 1944. The Ju 87G included two 37mm (1.46in) Flak 18 guns designed for knocking out tanks and armoured fighting vehicles.

Junkers Ju 87 G-1
Type: Two-seater dive-bomber
Length: 11.1m (36ft 5in)
Wingspan: 13.8m (45ft 3in)
Height: 4.24m (13ft 11in)
Weight: 6600kg (14,550lb) maximum take-off
Powerplant: 1 x 1044kW (1400hp) Junkers Jumo 211J piston engine
Maximum speed: 314km/h (195mph)
Range: 600km (373 miles)
Service ceiling: 8100m (26,575ft)
Armament: 2 x 3.7cm (1.46in) Bordkanone BK 3.7 auto-cannon; 1 x 7.92mm (0.31in) MG

BK 7.5 gun made the B-3 unstable and much less manoeuvrable, and there were problems with the cannon's reliability. Despite its undoubted tank-busting capability, production of the Hs 129 was terminated in 1943 in favour of more pressing projects such as the development of jet fighters.

Junkers Ju 87 G Stuka

The G series was the final development of what was by late 1942 a rather aged Ju 87 design that would soon struggle with the capabilities of the latest Allied aircraft to enter combat. The G series variants of the Stuka were dedicated ground-attack platforms intended primarily for close-support missions. This tactical role gave the old Ju 87 Stuka design a new, albeit temporary, lease of life. This design direction was prompted by the proven vulnerability to enemy fire of the existing ground-attack workhorse, the Hs 129B.

Production of the Ju 87 G-1 variant commenced in late 1942, and the

Ju 87 Gustav tank buster

Frontal view of a Ju 87 Stuka G-1 ground-attack aircraft; the size of the aircraft's large under-wing pod-mounted 3.7cm (1.46in) Bordkanone 3.7 'tank-busting' auto-cannon is clearly evident in this image.

first 20 completed airframes reached the Eastern Front in spring 1943. Both the G-1 and subsequent G-2 designs featured two 3.7cm (1.46in) Bordkanone BK 3.7 auto-cannon mounted in under-wing pods. The G-1 was a modification of the basic Stuka D-3 design, while the G-2 was a derivative of the D-5 model. Each BK 3.7 auto-cannon carried two six-round magazines of Armour-Piercing Composite Rigid (APCR) tungsten carbide cored rounds. These cannon turned the Ju 87 G into a potent tank-killer, although each aircraft had only 24 rounds to deliver before needing to be rearmed. It was this undoubted tank-killing capability that gave these aircraft their popular nickname, Kanonenvogel ('Cannon-birds').

During Operation 'Citadel', the titanic July 1943 armoured clash in the Kursk salient, only a handful of Model G Stuka tank-busters were available for combat; nevertheless, they secured a significant number of kills of Soviet armour in this target-rich, open-terrain environment. In the second half of 1943, moreover, Model Gs in the hands of highly experienced 'aces' such as Hans-Ulrich Rudel accounted for the destruction of hundreds of Soviet AFVs, as a string of Red Army offensives saw their massed armoured columns charge westward deep into Axis lines. Indeed, in October 1943, Rudel achieved his 100th confirmed destruction of a Soviet tank.

In early 1944, Ju 87 G-equipped ground-attack units, such as Schlachtgeschwader 2 (Ground-attack Wing 2), continued to inflict heavy losses on Soviet armoured vehicles; on 23 March 1944, for example, Rudel achieved his 200th confirmed enemy tank 'kill'. These Model G aircraft, also supported by heavily armed Stuka P-1s and P-2s, similarly inflicted heavy losses on the Soviet armour that surged rapidly westward during the successful summer 1944 Bagration offensive. With production stopped in 1944 and with heavy combat losses, only a few dozen Ju 87 Gs remained operational by spring 1945. Some of these surviving platforms mounted their last operations in late April 1945, trying to stem the mass of advancing Soviet armour that was then surrounding Berlin.

BOMBER AIRCRAFT

Luftwaffe bombers played a key role in the success achieved by German ground forces during the 1939–45 war, and also aided Kriegsmarine operations. Bomber missions included strategic air attack, striking key locations such as cities to break the enemy's will. Alternatively, Luftwaffe light/medium bombers conducted surprise counter-air campaigns, bombing enemy airfields and associated installations (as during the May 1940 Blitzkrieg in the West).

Having thus secured aerial superiority, Luftwaffe bombers could conduct missions supporting the ground offensive. These included 'close air support', where the Ju 87 Stuka dive-bomber played a key role, and 'interdiction' missions, where Luftwaffe light/medium bombers struck rail/road junctions or reserve forces. Below we examine the six most important German bombers: the Do 17 and 217 high-speed light bombers; the He 111 medium bomber; the He 177 long-range heavy bomber; the Ju 87 Stuka dive-bomber; and the Ju 88 high-speed medium bomber.

Dornier Do 17 high-speed light bomber

Nicknamed the 'Flying Pencil' after its long and thin fuselage, the Dornier Do 17 was a high-speed light bomber. Visually, the aircraft featured a set of shoulder wings that ran across the top of the fuselage; it also had a twin tail-fin arrangement. A wide-viewed

Dornier Do 17 Z-2
Type: Four-seater bomber
Length: 15.79m (51ft 9in)
Wingspan: 18m (59ft)
Height: 4.56m (14ft 11in)
Weight: 9000kg (19,841lb) maximum take-off
Powerplant: 2 x 746kW (1000hp) Bramo 323P Fafnir nine-cylinder radial engine
Maximum speed: 425km/h (263mph)
Range: 1160km (720 miles)
Service ceiling: 8150m (26,740ft)
Armament: 6 x 7.92mm (0.31in) MG; 1000kg (2205lb) bombload

Do 17 Z-2
This Do 17 Z-2, platform 'U5+BH', served with the 1st Squadron, Kampfgeschwader ('Bomber Wing') 2 during the spring 1941 Axis invasion of Greece; note the formation insignia – a black eagle carrying a bomb on white background.

Bombing up
Using what appears to be a local-procured or self-improvised horse-drawn sled, ground crew replenish with bombs a Do 17Z then serving with the III Gruppe, Kampfgeschwader 3 'Blitz' in the Soviet Union, during winter 1941–42.

crew cockpit sat atop of the front upper fuselage while another cockpit nacelle filled the nose of the aircraft's fuselage. One engine was mounted high upon the leading edge of each wing. Designed in 1933–34, the first prototype aircraft (with just a single-fin tail), which was designated V1, made its maiden flight on 23 November 1934. During 1934–35, the V1 airframe, together with the other two extant twin tail-fin prototypes (V2 and V3), undertook extensive testing.

The first production model was the three-crew Do 17 E-1, which entered Luftwaffe service in 1937. This model was powered by two BMW VI 7.3D engines and carried a bomb payload of 250kg (550lb); an additional 500kg (1110lb) bomb could be carried on an external rack fitted under each wing. It could reach a top speed of 330km/h (205mph) in level flight and 500km/h (310mph) in a shallow dive. In terms of armament, the E-1 mounted two 7.92mm (0.31in) MG 15 machine guns, located in the rear cockpit position and the forward fuselage bottom position.

Another early production variant was the F-1 long-range reconnaissance aircraft, which entered service in 1938. In September 1938, the Luftwaffe had 578 Do 17s on its strength, these mainly

Other Do 17 variants	
J-1, J-2	E-2s fitted with BMW 132F engines
M	Medium bomber variant powered by Bramo 323 engine (200 built)
P-1	Reconnaissance model powered by BMW 132N engines (240 P models built)
P-2	A P-1 that incorporated under-wing external bomb racks
U-1	Pathfinder variant with five-man crew
Z-7	Night-fighter variant

being Model E-1s and F-1s, together with a few dozen Model M bombers and Type P reconnaissance variants.

The most numerous production variant was the five-crew Do 17 Z series. These entered service in early 1940, having incorporated lessons learned during the Spanish Civil War. The Z series aircraft had a redesigned nose and the crew cockpit lengthened to create a rear gunner position. The Z-2 model was the principal design produced in this series. Powered by the newly modified 746kW (1000hp) Bramo 323P-1 engine, the Z-2 carried a significantly larger bomb payload of 1000kg (2205lb). This version mounted between four and eight 20mm (0.79in), 13mm (0.5in) and 7.92mm (0.31in) machine guns, with two of these firing out of the upper sides of the crew cockpit pod.

The Z-2 first saw service in the May–June 1940 German invasion of the West, with 338 airframes participating in combat operations. The Do 17 Z also participated in the summer 1940 Battle of Britain, during which it suffered high losses at the hands of Allied fighters. Small numbers of Do 17s participated in the bombing operations undertaken to support the summer 1941 Axis invasion of the Soviet Union. Subsequently, these aircraft were gradually relegated to secondary duties as the more effective Do 217 bomber came on strength. By the time that Do 17 production ended in late 1940 (with resources shifting to Do 217 bomber manufacture), some 2139 aircraft had been produced.

Dornier Do 217 bomber

The Dornier Do 217 medium bomber was a longer-ranged improvement of the original Do 17. A shoulder-winged cantilever monoplane, the Do 217 featured a wider wingspan and a twin-tail arrangement; it could also carry a larger bomb payload than the Do 17. It was designed with dive-bombing as well as level strike capability. Seven slightly differing prototypes were subjected to a lengthy evaluation programme during 1938–39.

The E-2 variant became the first to enter mass production in March 1942. This model was powered by two BMW 801L air-cooled radial engines, one located under each wing, which enabled it to reach a top speed of 535km/h (332mph). In terms of defensive armament, the E-2 sported one 15mm (0.59in) MG151 cannon and five MG 15 machine guns. Subsequently, the E-3, which featured an up-armoured cockpit, entered service, as did the E-4, which lacked dive brakes. The final E series model, the E-5, was an anti-shipping variant that had external racks to carry the Hs 293 glide bomb.

Some 220 examples were manufactured of the next Do 217 variant to enter service, the K-1. This design was optimized for a night-fighting role. It had a modified defensive armament arrangement that included three twinned MG 81Zs. The K-1 was the first Do 217 design to incorporate the GM-1 nitrous-oxide engine injection system, which boosted the aircraft's maximum high-altitude speed by up to 84km/h (52mph). Some 51 K-1s were converted to the K-2 anti-shipping aircraft, fitted with external rails to carry the Fritz-X guided glide bomb (the world's first 'precision-guided munition').

The next major production series, the Do 217 M, was typically powered by the 1287kW (1726hp) Daimler-Benz DB 603A-1 liquid-cooled, 12-cylinder inverted V-12 engine. Entering service from 1942, most M series designs were intended for night-bombing operations. Subsequently, Dornier developed the J and N series night-fighter variants, of which 356 examples were produced. By the time production ended in December 1943, Dornier had produced 1451 Do 217s of all types.

Do 217 N-2/R22

The rare Do 217 N-2/R22 was a standard night fighter modified to also mount (in addition to the normal eight nose machine guns) four upward-firing 70° obliquely-sloping 'Schrage Musik' ('Slanted Music') 20mm (0.79in) MG 151/20 cannon in the top of the dorsal fuselage.

Dornier Do 217 N-2/R22

Type: Four-seater bomber
Length: 18.9m (62ft)
Wingspan: 19m (62ft 4in)
Height: 5m (16ft 5in)
Weight: 13,700kg (30,202lb) maximum take-off
Powerplant: 2 x 1380kW (1850hp) DB 603A engine
Maximum speed: 500km/h (310mph)
Range: 1755km (1090 miles)
Service ceiling: 8400m (27,560ft)
Armament: 4 x 20mm (0.79in) cannon; 4 x 7.92mm (0.31in) MG fixed forwards; 4 x 20mm (0.79in) cannon firing obliquely upwards

He 111 H-16s, Russian Front
A squadron of He 111 H-16 aircraft in flight on the Eastern Front; powered by two 1000kW (1340hp) Jumo 211 F-2 engines, this variant had a top speed of 434 km/h (270 mph).

Heinkel He 111 medium bomber

A twin-engined monoplane with forward-positioned, low-slung wings, the Heinkel He 111 was one of the most important aircraft employed by the Luftwaffe during the 1939–45 war. Visually, the later variants of the aircraft could be identified by their large glazed fuselage nose and by the gunner's canopy in the upper fuselage above the wings. The first two He 111 prototypes were put through exacting trials in 1935–36.

The B-1 variant was the first standard production model and entered service in early 1937. Powered by two DB600 engines, the B-1 could carry a 1500kg (3310lb) bomb payload and obtained a maximum speed of 344km/h (215mph). Subsequently, the He 111 D-1 entered limited production in late 1937.

Introduced in 1938, the next sub-type, the E-1, was powered by two 693kW (930hp) Jumo 211-A engines; the increased thrust enabled this variant to carry a larger bombload of some 2000kg (4410lb). The next major modification, incorporated into F series aircraft, was the introduction of straight-edged wings rather than elliptical ones. Introduced in late 1938, the He 111 P series incorporated the up-rated 809kW (1085hp) DB601A liquid-cooled inverted V-12 engine. This increased thrust increased the Model P's top speed to 475km/h (295mph). This model was easily recognized due to the introduction of a completely glazed fuselage nose.

Introduced in 1939–41, the early He 111 H series (H-1 to H-10) was the most widely produced sub-type. At the start of the 1939–45 war, the Luftwaffe had 705 He 111s, including 400 Model Ps and numerous Model Fs and H-1s; these played a significant role

He 111 H-3

This Heinkel He 111 H-3 designated '1H+MM' served with Kampfgeschwader 26 in the Mediterranean during 1943-44 and had a white vertical theatre stripe on the rear fuselage; the lion insignia of KG26 is just visible on the nose, partly obscured by the engine.

Heinkel He 111 H-3
Type: Four-seater bomber
Length: 16.4m (53ft 9in)
Wingspan: 22.6m (74ft 2in)
Height: 4m (13ft 1in)
Weight: 14,000kg (30,864lb) maximum take-off
Powerplant: 2 x 895kW (1200hp) Junkers Jumo 211D 12-cylinder engine
Maximum speed: 415km/h (258mph)
Range: 1200km (745 miles)
Service ceiling: 7800m (25,590ft)
Armament: up to 7 × 7.92mm (0.31in) MG 15 or MG 81 machine guns; 1 × 20mm (0.79in) MG FF cannon; 2000kg (4400lb) bombload

in the autumn 1939 invasion of Poland. Similarly, more than 900 He 111s undertook strategic and operational-level bombing missions during the May 1940 German Blitzkrieg campaign in the West; in particular, He 111s undertook the infamous 'terror blitz' on Rotterdam on 14 May 1940. Some 450 recently manufactured H-1, H-2 and H-3s also spearheaded the summer 1940 Battle of Britain, but suffered losses of 242 airframes.

The later up-armoured H series (H-11 to H-20) entered service from summer 1942. These featured up-armoured cockpits and enhanced defensive armament, including three high rate-of-fire twin 7.92mm (0.31in) MG 81Z machine guns. These later H series aircraft undertook numerous medium-range bombing missions to support the ground operations undertaken on the Eastern Front; for example, bombing the Soviet railway network or production factories. At

Heinkel He 111 annual production

Year	Annual production
1935–39	1260
1940	930
1941	950
1942	1337
1943	1405
1944	756
Total	6638
1945	2800
Total	33,984

both Demyansk (winter 1941–42) and Stalingrad (winter 1942–43), the He 111 was used as an improvised transport platform. By 1944, many He 111-equipped units had been relegated to second-line duties. In total, the six German factories involved in He 111 manufacture completed some 6638 examples.

Bomb loading an He 177
German ground crew re-load with bombs this He 177 A-5 aircraft, which apparently bears in white on its front fuselage the name 'Helga'.

Heinkel He 177 Greif ('Griffin') heavy bomber

The Heinkel He 177 was the only Luftwaffe long-range heavy bomber to see service in the war with a capability that matched the Allied Lancaster or B-17 strategic bombers. The high-speed twin-engined aircraft was designed to be capable of shallow-glide bombing as well as level bombing. The first prototype two-engined He 177, designated V1, flew in late 1939; eight slightly different prototypes (V2-V8) were subjected to extensive testing in 1940–41.

Manufacture of the main production version, the He 177 A-1, commenced in January 1942. Over the next 12 months, 260 units were completed. Like the prototypes, the A series aircraft featured a distinctive 'fish bowl' glazed fuselage nose. The standard A-1 was powered by two DB606 'power system' engines – namely, each engine actually comprised two DB 601 engines mounted side by side in one nacelle that drove a single propeller. Initial operational service showed that the tight nacelle around these twinned engines seriously hampered maintenance and the engines soon earned a fearsome reputation for bursting into flames.

Manufacture of the modified A-3 variant commenced in November 1942. This design was powered by slightly modified DB 610 'engine systems' in a redesigned nacelle. The A-3 aircraft also featured an elongated fuselage that helped to reduce the degree of instability experienced in the A-1 design in terms of pitching and yawing. Later A-3s also featured a redesigned enlarged rear gun position.

The main production variant, the modified A-5, entered construction in December 1943. This variant incorporated strengthened wings, an elongated fuselage, and shortened undercarriage legs. The A-5 could obtain a top speed of 565km/h (351mph) and a maximum operating ceiling of 8000m (26,246ft). The design mounted two 20mm (0.79in) MG 151 cannon, together with four 13mm

He 177 A-5/R2 Grief

This He 177 A-5/R2 served with the 4th Squadron, Kampfgeschwader 40 during 1944; the up-armed A-5/R2 variant bristled with no fewer than seven or eight machine guns and cannon.

He 177A-5/R2 Grief
Type: Six-seater bomber
Length: 22m (72ft 2in)
Wingspan: 31.44m (103ft 2in)
Height: 6.67m (21ft 10in)
Weight: 32,000kg (70,548lb) maximum take-off
Powerplant: 2 × Daimler-Benz DB 610, each one created from a twinned-pair of Daimler-Benz DB 605 inverted V12 engines, 2900 PS (2133kW) each
Maximum speed: 565km/h (351mph)
Range: 1540km (957 miles)
Service ceiling: 8000m (26,246ft)
Armament: 3 × 7.92mm (0.31in) MG 15 or MG 81 machine guns; 3 x 13mm (0.5in) MGs; 2 × 20mm (0.79in) MG FF cannon; 7200kg (15,873lb) bombload

(0.5in) MG 131 and one 7.92mm (0.31in) MG 81 machine guns. This design could carry internally a 6000kg (13,227lb) bombload together with an external load of 7200kg (15,873lb).

In 1942–44, Heinkel developed a four-engined version of the Greif, the Model B. Three prototypes powered by four DB603 engines separately mounted on lengthened wings were flown during 1943–44. The introduction of the July 1944 'Emergency Fighter Programme', however, stopped any further work on the He 177 B project. Some 18 He 177 A-3s joined 456 other bombers in executing the 'Baby Blitz', the

He 177 production figures		
Model	**Number produced**	**Period**
A-0	15	Late 1941
A-1	260	January 1942–January 1943
A-3	1234	November 1942 –June 1944
A-5	701	December 1943–August 1944
Total	2210	

early 1944 German strategic bombing campaign against London; due to their fast extraction speeds, only two A-3s were lost to enemy fire (out of a total of 329 German losses), although engines fires still occurred. Thereafter, crippling fuel shortages effectively grounded the squadrons equipped with the Greif for the remainder of the war.

Junkers Ju 87 bomber

The all-metal Ju 87 monoplane dive-bomber/ground-attack aircraft was invariably known as the Stuka (from Sturzkampfflugzeug, 'dive-bomber'). The Stuka incorporated several distinctive features. First, it sported bent inverted gull-wings – wings that slanted down from the fuselage for the first third of the wing length and then inclined towards the tip. Another distinctive feature was the large spatted fairings above the fixed landing wheels, designed to reduce drag. The aircraft also featured a large glazed crew gondola and a conventional single-tail configuration. The story of the final Stuka series, the Model G ground-attack variant, has already been recounted.

The developmental origins of the two-crew Ju 87 hark back to experimental Junkers designs produced during 1927–34. In 1935–37, seven prototype Ju 87 designs and 20 Model A-0 pre-production airframes were subjected to extensive tests. Subsequently, the Ju 87 B series entered mass production in 1937. This model was powered by a single, nose-mounted, 882kW (1184hp) Junkers Jumo 211D

Stukas patrol over the Mediterranean

Two Ju 87 R-2 long-range variants fly over the Mediterranean coast; this variant of the B-2, powered by the Jumo 211-A engine, carried a 300-litre fuel drop-tank under each wing.

Ju 87 D-1/Trop, North Africa, 1942

In addition to augmented defensive firepower and significantly increased bomb-load, the Ju 87 D-1 incorporated two coolant radiators under the inner wings and the oil cooler moved to the under-nose 'chin' position.

Junker Ju 87 D-1
Type: Ground-attack aircraft
Length: 11.5m (37ft 9in)
Wingspan: 13.8m (45ft 3in)
Height: 3.9m (12ft 9in)
Weight: 6600kg (14,551lb) maximum take-off
Powerplant: 1 x 1044kW (1400hp) Junkers Jumo 211J-1 inverted-Vee piston engine
Maximum speed: 410km/h (255mph)
Range: 1535km (954 miles)
Service ceiling: 6100m (20,015ft)
Crew: 2
Armament: 3 x 7.92mm (0.31in) machine guns plus bombload of up to 1800kg (3968lb)

inverted V-12 engine. The two-crew Model B-1 aircraft could carry a bomb payload of 500kg (1100lb). This variant also featured a redesigned fuselage and landing gear. When the 1939–45 war commenced, the Luftwaffe had 336 Ju 87 Bs in service.

The Ju 87 D variant entered mass production in February 1940. This variant incorporated an improved cockpit and a twin-barrelled MG 81Z machine gun in the rear defensive position. Powered by the up-rated 1045kW (1402hp) Jumo 211J engine, the Model D could carry a payload of 1800kg (4000lb). The final variant, the effective ground-attack Model G variant, entered service in April 1943. It mounted two 37mm (1.46in) Bordkanone automatic anti-tank cannon, one under each wing. The Ju 87 G proved an effective tank-killer; leading German Stuka ace Hans-Ulrich Rudel claimed 519 enemy tanks destroyed. In late 1943, however, the Luftwaffe decided that the Stuka was too vulnerable to survive air combat in the East and production was phased out. Between 1935 and early 1944, four German factories manufactured some 6507 Ju 57s.

In terms of operational service, during the early years of successful Blitzkrieg (1939–41), the Stuka performed effectively in the close air support role, providing immediate fire support for the contact battle on the ground. It had sirens on its landing gear legs that wailed as the aircraft dived, which added a psychological effect to its lethality when attacking ground troops. In the early phase of the 1939–45 war, the Stuka also undertook anti-shipping strikes in coastal waters. However, being slow and lacking in manoeuvrability, the Stuka needed a strong fighter escort to operate effectively. On the Eastern Front from spring 1943 onwards, the final Ju 87 variant, the Model G, was used in a ground-attack role engaging Soviet AFVs; for example, during the summer 1944 Soviet Bagration offensive. Ju 87 G aircraft mounted their last operations in late April 1945, trying to stem the mass of advancing Soviet armour that was surrounding Berlin.

Junkers Ju 88 bomber

The Ju 88 was a two-engined monoplane with a conventional single tail fin and low-mounted forward-positioned wings; early versions had a distinctive fully glazed fuselage nose. Originally designed as a high-speed light bomber and then a medium bomber, this versatile design was modified to produce fighter-bomber, night-heavy fighter and reconnaissance variants. Five different Ju 88 prototype high-speed bomber designs were evaluated during lengthy trials in 1936–39. The last two designs differed from the preceding three by being less fast medium bomber designs optimized to also enable a dive-bombing capability; they featured strengthened wings, dive brakes, and were manned by four (rather than three) crew-members.

The first model to enter service, the A-1, did so in 1939, but numerous

Junkers Ju 88 A-5
Type: Three-seater bomber
Length: 15.58m (51ft 1in)
Wingspan: 20m (65ft 7in)
Height: 4.85m (15ft 11in)
Weight: 14,000kg (30,864lb) maximum take-off
Powerplant: 2 x 1000kW (1340hp) Junkers Jumo 211 12 cylinder engines
Maximum speed: 433km/h (269mph)
Range: 2250km (1398 miles)
Service ceiling: 8200m (26,900ft)
Crew: 3
Armament: 6 x 7.92mm (0.31in) MG 81; max bomb load 3000kg (6614lb)

Ju 88 A-5
Serving with Kampfgeschwader 30, this Ju 88 A-5 bomber was a converted existing A-series platform with lengthened wings and powered by two up-rated Jumo 211 H-1 engines.

Ju 88 G-1

This Ju 88 G-1 was operated by 7./NJG 2. After its crew became lost on the night of 12–13 July 1944, it landed by accident at RAF Woodbridge in the United Kingdom. It provided the British with valuable intelligence.

technical issues meant that initial production was desultory. Model A aircraft participated in the 1940 Western campaign, particularly with dive-bombing strikes, as well as in the Battle of Britain.

The next significant variant developed was the Ju 88 C fighter-bomber/heavy night-fighter variant, which was introduced in 1941. The first 20 C-1s machines were converted from A-1s in mid-1940. Unlike its predecessors, C series aircraft had an all-metal nose that housed one 20mm (0.79in) MG FF auto-cannon and three 7.92mm (0.31in) MG 17 machine guns. From 1942, many Ju 88 C-1/C3s were based in France and undertook anti-shipping and escort duties in the Bay of Biscay. During 1942–44, these Ju 88s achieved 108 confirmed aerial combat 'kills', but lost 117 platforms in the process.

The final night-fighter variants were the R-1 and R-2, which were powered, respectively, by two 1147kW (1539hp) BMW 801L air-cooled 14-cylinder engines or two BMW 801G-2 engines.

The G series aircraft were the first Ju 88 night-fighters built from scratch for this role and featured a fundamentally redesigned fuselage. They sported a radar system and aerials – usually the FuG 220 with eight-dipole aerials in the nose. The G-6, for example, was powered by two Jumo 213A inline-V12 engines; its armament included two 70-degree slanted 'Jazz Music' 20mm (0.79in) MG 151/20 cannons located in the upper fuselage.

The Ju 88 P was a specialized model optimized for both ground-attack and aerial anti-bomber missions that featured the all-metal nose seen in the C series airframes. The 40 Model P-1 aircraft manufactured mounted a 7.5cm (2.95in) Bordkanone [BK] semi-automatic gun in a ventral

NIGHT-FIGHTER VARIANT

The principal sub-type in the C series was the C-6 night-fighter, of which 900 aircraft were manufactured in 1942–43. C-6s were recognizable by the prominent Matratze 32-dipole antennae that projected from the nose to support their FuG-202 Lichtenstein BC low-UHF band interception radar. Later C-6s, however, featured the large eight-dipole Hirschgeweih ('stag's antlers') aerials for the FuG-220 VHF-band radar they carried.

under-fuselage pod. The P-2 and P-3s platforms, moreover, sported a twin 3.7cm (1.46in) BK automatic cannon in its ventral pod. The final design in this series, the P-4, mounted a single ventral-pod-housed 5cm (1.96in) BK auto-cannon.

JET AIRCRAFT

During the second half of the war, the Luftwaffe hoped that a new generation of high-speed jet aircraft would enable Germany to wrest back the strategic initiative in the battle for the skies above Europe. During 1939–44, the Luftwaffe, through the allocation of extensive resources from the hard-pressed war economy, developed operationally viable turbojet engines, rocket motors and appropriate airframes.

This section examines the five most important jet aircraft: the Messerschmitt Me 262 and Heinkel He 162 turbo-jet aircraft; the Arado Ar 234 turbojet bomber/reconnaissance aeroplane; and the Messerschmitt Me 163 and Bachem Ba 349 rocket-powered interceptors. Only small numbers of these designs became operational before the war's end. Their effectiveness was hampered by shortages of fuel, trained pilots and ground crews, as well as by a lack of suitable runways and airfield infrastructure; the astute Allied tactic of engaging these jets during take-off and landing also hampered these missions. Consequently, despite the vast amount of resources poured into developing these jet aircraft, they downed only around 700 Allied aircraft.

Messerschmitt Me 262 jet fighter/ fighter-bomber

Visually, the Messerschmitt Me 262 featured a sleek, streamlined fuselage and a pair of swept wings, each with a large cylindrical under-wing nacelle that housed the turbojet engine. The first prototype Me 262 airframe made its maiden flight in April 1941, but powered by an ordinary piston engine; the development of the earmarked

Me 262 A-1a, summer 1944, Lager-Lechfeld
Flown by pilot Fritz Müeller, this Me 262 A-1a (WNr 170059) served with 'Testing Unit' 262 (Eprobungskommando 262) at either the Lechfeld or Leipheim airbases in Bavaria, south of Augsburg, during summer 1944.

Junkers Jumo 004 turbojet engines had fallen behind schedule. These turbojets were only installed in the third prototype aircraft, V3, for its first flight on 18 July 1942.

The Me 262 had originally been designed as a fighter, but during summer 1943 Hitler insisted that it also be developed as a light ground-attack fighter-bomber aircraft. The Me 262-A-2a Sturmvogel ('Storm Petrel') fighter-bomber variant carried up to 500kg (1102lb) of bombs in addition to the four 3cm (1.18in) MK 108 nose cannon mounted by both the fighter-bomber and fighter variants. Later Me 262s of both varieties also mounted 24 R4M rockets.

The fighter version, the Me 262 A-1a Schwalbe ('Swallow'), was significantly delayed by the teething problems encountered with the Jumo 004 engines. Consequently, this fighter variant did not enter Luftwaffe service until late 1944. General production commenced in June 1944, when 28 aircraft were produced, plus another 79 over the ensuing two months. Even with this delayed service entrance, the Me 262 was probably the most advanced aeroplane operational at that moment. This fighter variant, powered by two Jumo 004B engines, was armed with four 3cm (1.18in) MK 108 cannon in the nose. With its powerful turbojets, the Me 262 A-1a fighter could obtain a maximum speed of 870km/h (541mph). Early Jumo engines proved to have short service lives before extensive refitting was required. The Schwalbe carried some 2000 litres (440 gallons) of fuel, but given the design's ferocious fuel consumption this was sufficient for just 60–80 minutes' worth of flight.

The 16 confirmed 'kills' of Me 262 ace Heinz

Aircraft type	Total kills
B-24 bomber	2
B-26 bomber	3
P-47 fighter	7
P-51 fighter	4

Me 262 A-1a/U4 with 50mm (1.96in) Mauser cannon
The rare A-1a/U4 'bomber-killer', of which just two aircraft were converted during spring 1944 at Lechfeld airbase; these platforms mounted in the nose a long-barrelled 5cm Mk 214a auto-cannon, a derivative of the Pak 38.

Me 262 A-1a

This Me 262 A-1a aircraft 'Yellow 8' (WkNr 112385) of the 3rd Squadron, Jagdgeschwader 7, was captured by American ground forces on 15 April at an airbase near Stendhal.

Messerschmitt Me 262 A-1a
Type: Single-seat fighter
Length: 10.61m (34ft 9in)
Wingspan: 12.5m (41ft)
Height: 3.83m (12ft 7in)
Weight: 6775kg (14,936lb) maximum take-off
Powerplant: 2 x 8.8kN (1890lb) Junkers Jumo 004B-1 turbojets
Maximum speed: 870km/h (541mph)
Range: 845km (525 miles)
Service ceiling: 11,000m (36,090ft)
Armament: 4 x 30mm (1.18in) cannon

The Schwalbe first undertook combat missions in August 1944, serving in the Kommando Nowotny; during that month, this unit claimed the destruction of 18 Allied aircraft for the loss of six Me 262s. The Sturmvogel first saw combat in a ground-attack role during October 1944 with Kampfgeschwader 51.

Increasing numbers of Me 262s began reaching operational airfields in spring 1945. Indeed, despite facing huge challenges, the hard-pressed German war economy managed to produce 1400 Me-262s before the war's end. However, due to significant transportation problems, only 300 Me 262s actually undertook operations against the Allies. During March, these higher numbers available for combat permitted mass attacks for the first time. On 18 March 1945, for example, 37 Me 262s attacked a formation of 1221 bombers, that was protected by 627 fighters; the Me 262s downed 12 bombers for the loss of three planes.

The overall effectiveness of the Me 262, however, was hampered by the clever tactical responses of the Allies;

Me 262 B-1a/U1 night fighter

This Me 262B-1a/U1 Werk/nr 111980 was assigned to 10./NGJ 11, better known as Kommando Welter, as a night fighter. 'Red 12' was operated from Burg bei Magdeburg until May 1945.

Me 613 B-1a, Bad Zwischenahn, summer 1944

A Me 613 B-1a takes off from Bad Zwischenahn airbase in Oldenburg in north-western Germany, summer 1944; this variant had slightly-modified aerodynamically-improved wing design that gave the platform a higher critical Mach number.

their fighters engaged Me 262s either when they were taking off or landing, when the latter's technical superiority was minimized. Despite this, these 300 operational Me 262s proved highly effective, with their pilots claiming to have downed 542 Allied aircraft; at least 120 Me 262s were lost in the process.

Messerschmitt Me 163 Komet ('Comet') rocket-powered fighter

The developmental origins of the Messerschmitt Me 163 can be traced back to 1940; indeed, during 1941, the first prototype Komet airframe, V1, flew as an unpowered glider. Subsequently, V1 was fitted with the potent Walter HWK rocket motor, which enabled it to reach an impressive top speed of 960km/h (596 mph). On 2 October 1941, during test flights at Peenemünde, the fourth Komet prototype (V4) became the first aircraft in history to break through the 1000km/h (620mph) barrier.

After the construction of five prototypes (V1–5) in 1941, eight pre-production Me 163 A-0 aircraft were manufactured for further trials. The Model A incorporated as its landing gear a two-wheeled dolly arrangement. The testing squadron EK 16 conducted extensive trial flights on these eight Model A-0s in the second half of 1943. These tests revealed that the design was not optimized for mass production, so a slightly modified, easier to produce variant, the Me 163 B, was developed.

The first Me 163 Bs produced in spring 1944 were 30 B-0 pre-production models, which were armed with four 2cm (0.79in) MK 108 cannons. In total, some 400 production Model B aircraft were manufactured before the war's end. The Me 163 B plane incorporated a different powerplant: the Walter 109-509A-2 rocket motor. This device utilized a combination of three parts of the highly combustible T-Stoff peroxide oxidizer to one part of C-Stoff methanol-hydrazine fuel. These substances had to be held in separate fuel tanks to prevent spontaneous combustion; the T-Stoff tank was positioned behind the pilot's head in the upper central fuselage, and the C-Stoff tank further back in the upper fuselage just before the tail. Unfortunately, the Me 163 B could only carry sufficient oxidizer and propellant to sustain a little over eight minutes of powered flight, so this technically impressive airframe was limited tactically to a point defence interceptor. On 6 July 1944, a Model B Komet set an unofficial world speed record of 1130km/h (702mph), a speed not exceeded until 1947.

During January–April 1944, the unit EK 16, now located in northwestern Germany, near Oldenburg, extensively tested the first Model B-0 aircraft. In May 1944, a prototype B-0 flew this variant's first combat missions. Attacking usually in pairs, the Komet pilot would dive down at extreme speed into an Allied bomber formation firing its cannon, before climbing back up, repeating the tactic, then flying back to base. In summer 1944, Jagdgeschwader (Fighter Wing) 400 became the first operational unit to employ the Komet. Stationed near Leipzig, its 42 operational Model B1-a interceptor aircraft shot down their first American B-17 Flying Fortress bombers on 24 August 1944.

Chronic shortages of aircraft, fuel, adequately trained pilots and maintenance personnel, together with frequent Allied attacks on its airbase at Brandis, all combined to limit the number of effective missions flown by JG 400's Me 163s. In addition, due to its incredible performance, the aircraft was not easy to fly; targeting needed a high level of professional skill, a

Messerschmitt Me 163 B-1a
Type: Single-seat fighter
Length: 5.7m (18ft 8in)
Wingspan: 9.3m (30ft 6in)
Height: 2.5m (8ft 2in)
Weight: 4309kg (9500lb) maximum take-off
Powerplant: 1 × Helmut Walter Kiel Kommandogesellschaft HWK 109-509A-2 bi-propellant liquid-fuelled rocket motor, 14.71kN (3307lbf) thrust maximum
Maximum speed: 900km/h (559mph)
Range: 7.5 mins powered
Service ceiling: 12,000m (39,000ft)
Armament: 2 × 30mm (1.18in) Rheinmetall Borsig MK 108 cannon with 60rpg (B-1a)

Me 163 B-1a
The short, stubby profile of the Me 163 B-1a is evident here; note on the upper fuselage the small white 'T' and yellow 'C' markings to indicate where the separate T-Stoff and C-Toff fuel tanks were located.

commodity often in short supply at this late stage of the war. Taken as a whole, Komet combat sorties caused the loss of between just nine and 17 Allied aircraft, while Allied actions downed 10 Me 163s; many more aircraft were lost during training/combat accidents or else due to structural failures.

Heinkel He 162 Salamander jet fighter

In 1944, the Germans realized that, whatever their technical abilities, sophisticated jet fighter aircraft like the Me 262 and Me 163 consumed large amounts of precious resources and skilled technical labour capabilities and complex machinery. The hard-pressed German war economy could only produce these complex pieces of engineering in modest amounts over extended periods of time. There would never be sufficient Me 262s and Me 163s to wrest back the strategic initiative in the skies above the German Reich. Consequently, the High Command decided that they needed a simple, easily mass-produced, easily piloted, lightweight jet fighter; the July 1944 'Emergency Fighter Programme' held a competition to produce the best such design. During summer 1944, therefore, the Germans developed an unsophisticated jet interceptor that was both easily produced with semi-skilled workers, and cheap to manufacture in terms of resources. The ensuing design was designated the Heinkel He 162 Salamander, although it was known in common parlance as the Volksjäger ('the People's Fighter') or the Spatz ('Sparrow').

In terms of its external appearance, this aircraft was a small, round-nosed, streamlined airframe that featured short, stubby, forward-swept, trailing-edge wings set high in the fuselage; it also featured a wide and shallow 'H' tail design. The large jet engine, a single BMW 003E turbojet, was positioned on top of the fuselage just behind the pilot's canopy; the diameter of this

He 162 A-2, 3. Staffel/JG 1, May 1945

He 162 A-2 'White 23' (WkNr 120222) of the Headquarters of Jagdgeschwader 1 is finished in light green (RLM 82) on upper surfaces with white-blue (RLM 76) of lower surfaces.

He 162 A-2
The He 162 A-2 aircraft (WkNr 120074) of Jagdgeschwader 1 flown by pilot Erich Demuth from Leck airfield near the German-Danish border during late spring 1945. The tail fin records Demuth's 16 kills, but these were achieved in the previous, non-jet, aircraft that he flew.

Heinkel He 162A-2
Type: Single-seat fighter
Length: 9.05m (29ft 8in)
Wingspan: 7.2m (23ft 7in)
Height: 2.6m (8ft 6in)
Weight: 2800kg (6180lb) maximum take-off
Powerplant: 1 × BMW 003E-1 or E-2 (meant for dorsal fuselage attachment) axial flow turbojet, 7.85kN (1760lbf)
Maximum speed: 790km/h (491mph) sea level; 840km/h (522mph) at 6000m (19,680ft)
Range: 975km (606 miles)
Service ceiling: 12,000m (39,000ft)
Armament: 2 × 20mm (0.79in) MG 151/20 autocannon

broadly cylindrical engine nacelle was as large as that of the aircraft's fuselage. The lightweight plane, however, could only carry sufficient fuel for a 30-minute flight. The He 162 also featured a traditional retractable 'tricycle' undercarriage arrangement of one nose wheel and a wheel each in the outer lower fuselage.

After extremely rushed development work, manufacture of two He 162 prototypes commenced in October 1944. The first test flights were flown on 6 December 1944, an incredibly short time for this landmark point to be reached. These tests, however, revealed issues with instability in the wing tips and the rotting impact on wood of the highly acidic glue used. Despite these concerns being identified, production of two further prototypes, incorporating only minor modifications, commenced anyway. A key stability-enhancing element incorporated was drooped wingtips. These two planes undertook further testing during mid-January 1945.

The A-2 production fighter variant entered manufacture in January 1945; that month, the first 46 aircraft were rushed to test unit EK 162 at Rechlin for further testing. Production facilities had been hurriedly set up at three locations: at Salzburg, and in the Hinterbrühl and Dora-Mittelwerke underground slave-labour facilities. The Germans set a highly ambitious monthly production target of 1000 units per month; however, only 300 were completed and a mere 50 reached combat units.

The He 162 A-2 production variant mounted two 20mm (0.79in) MG151/20 cannons in the fuselage nose. Within weeks, many of the aircraft at Rechlin had been delivered to I/Jagdgeschwader 1, based at Parchim near Berlin; this became the first operational unit to be equipped with the He 162 in late March. Just as this unit was to commence operations, its capabilities were severely disrupted when, on 7 April, 134 B-17 Flying Fortresses inflicted severe damage on the Parchim airbase.

During the ensuing six days, I/JG1 relocated to Leck airfield, close to the Danish border in western Schleswig-Holstein. II/JG1 also began training with He 162s at Rostock. In the second half of April and early May, JG1 undertook dozens of sorties, during which its pilots shot down five Allied aircraft. In the process, Allied fire downed two Salamanders, while a further 11 were lost due to structural failures, engine flame-outs, or accidents. Postwar Allied testing confirmed that, in the hands of a well-trained pilot, and with expert maintenance, the He 162 would have been a highly effective interceptor.

Bachem Ba 349 Natter ('Viper') rocket interceptor

Designed in 1944, the Bachem Ba-349 Natter ('Viper') was one of the least well known of the jet aircraft developed by Germany during the war's final stages. The Natter was the world's first vertical take-off rocket-powered interceptor aircraft. The single-seater point-defence aircraft was designed to use wooden plates and an armoured cockpit. The design was so simplistic that it could be manufactured by semi-skilled labour using simple machine tools. Visually, the Natter had a cylindrical-shaped fuselage with short stubby wings and a T-shaped tail.

Bachem Ba 349B-1
The simplistic, easily-manufactured design of the Bachem Natter is evident here. The two Schmidding SG-34 booster rockets per side can be seen on the rear fuselage in front of the large T-shaped tail structure.

Bachem Ba 349B-1
Type: Single-seat fighter
Length: 6m (19ft 8in)
Wingspan: 4m (13ft 1in)
Height: 2.25m (7ft 5in)
Weight: 2232kg (4921lb) maximum take-off
Powerplant: 1 × Walter HWK 109-509C-1 bi-fuel rocket motor, 11.2kN (2500lbf) thrust Hauptofen main chamber; 2.9kN (652lbf) Marschofen auxiliary chamber
Maximum speed: 1000km/h (621mph) at 5000m (16,404ft)
Range: 60km (37 miles) after climb at 3000m (9843ft)
Service ceiling: 12,000m (39,000ft)
Armament: 24 × 73mm (2.874in) Henschel Hs 297 Föhn rocket shells

Powered by its Walter HKW 109-509C-1 bi-fuel rocket motor, plus four Schmidding SG-34 solid fuel booster rockets, the Natter rocketed vertically on autopilot at an astonishing speed of 1000km/h (621mph) to a height of 9000m (29,525ft). At this point, the pilot took control of the aircraft and flew it until he had located an enemy bomber formation. Flying at 800km/h (497mph), the pilot had just four minutes' worth of fuel to locate the enemy aircraft. Having approached to within 800m (2624ft) of the formation, the pilot fired the aircraft's 24 nose-mounted 73mm (2.87in) Henschel Hs 297 Föhn missiles. With the interceptor's fuel tanks now empty, the pilot glided it down to a height of 3000m (9842ft). At this point, he ejected and parachuted to safety as the 'Viper' crashed to earth.

The first full Natter test flight occurred on 1 March 1945. By the end of the month, 10 prototype Ba 349s stood ready for action at Kirchheim near Stuttgart. Unfortunately, no Allied bomber formations passed nearby during the few days that the battery was operational. In early April, the approach of American Army spearheads forced the Germans to hastily withdraw and then destroy all these 10 'Vipers'.

Arado Ar 234 Blitz ('Lightning') jet reconnaissance aircraft/ bomber

First flown in combat in June 1944, the Arado Ar 234 Blitz ('Lightning') became the world's first operational jet reconnaissance aircraft and bomber. The aircraft featured a smooth aerodynamic design and was powered by two Junkers Jumo 004-B turbojet engines. This

Ar 234s, Ardennes Offensive, January 1945
A group of Ar 234 bombers from Kampfgeschwader 76 on an airfield during the German Ardennes counter-offensive. On 26 December 1944, these planes undertook missions that struck the American forces counter-attacking the German advance around Bastogne, Belgium.

powerful propulsion system enabled the Ar 234 to obtain an impressive top speed of 735km/h (456mph), which enabled it typically to operate with impunity from slower Allied piston-engined Allied fighters.

Ar 234 development work commenced in 1940 on a high-winged monoplane powered by a turbojet engine that was mounted in a nacelle under each wing. The design sported two downward-looking reconnaissance cameras mounted in the rear fuselage. Unusually, the design eschewed a conventional landing gear in favour of retractable skids mounted beneath the fuselage. The aircraft would take off sitting on a wheeled trolley that was jettisoned as it left the runway.

Although Arado completed the Ar 234 V1 prototype airframe in late 1942, delays in the availability of Jumo engines slowed development work, as did the higher priority given to the development of the Me 262 jet aircraft. Consequently, the first flight of the prototype Ar 234 V1 recce aircraft did not occur until 30 July 1943. Six other recce prototypes were constructed during 1943–44. One of these, the V5 prototype, flew the first Ar 234 tactical mission on 6 August 1944 by mounting a reconnaissance sortie over the Allied beachhead in Normandy, France. These seven aircraft made numerous successful reconnaissance flights, during which they were never successfully engaged by Allied interceptors.

In the wake of these successful flights, the Air Ministry directed Arado to incorporate both bombing capability and a conventional tricycle landing gear into the aircraft. Arado enlarged the fuselage to accommodate the new landing gear and added a semi-recessed bomb bay under the fuselage. The Ar 234 B-2, the bomber variant first flown in March 1944, was the first sub-model to be mass-produced by the new Luftwaffe airfield factory at

KAMPFGESCHWADER 76

The deteriorating war situation, when coupled with disrupted transportation and shortages of fuel, ensured that just one Luftwaffe unit was equipped with operational Ar 234 bombers before Germany's surrender. This unit was KG 76 (Kampfgeschwader or 'Bomber Wing' 76), which flew its first Ar 234 bombing missions during the December 1944 Ardennes counter-offensive.

These aircraft flew in small groups of 10–20 aircraft, each carrying a single 500kg (1100lb) bomb. During mid-March 1945, the unit mounted desperate attacks against the Ludendorff railway bridge at Remagen over the River Rhine that the Allies had captured intact; these attacks did not destroy the bridge and the unit lost five aircraft to intense Allied anti-aircraft fire.

From late March until the war's end, lack of fuel, spare parts and pilots prevented KG 76 from flying more than a handful of sorties. The unit conducted its last missions against the Soviet forces encircling Berlin during the final days of April. During the first week of May, nine of the unit's surviving aircraft flew to airfields in German-occupied Norway, while the other four were destroyed at their base in Schleswig-Holstein to prevent their capture.

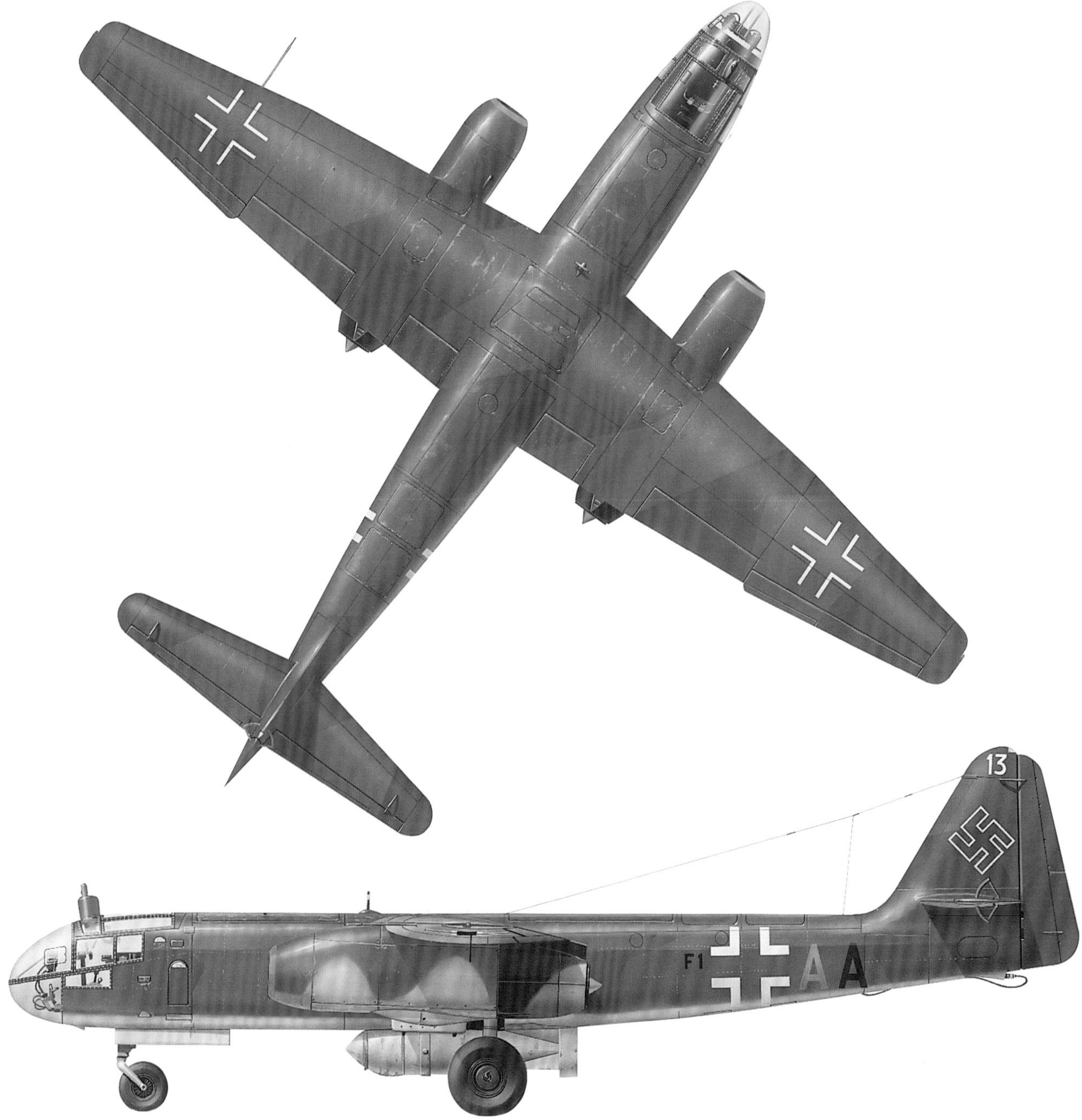

Arado Ar 234 B-2
Type: Single-seat bomber
Length: 12.64m (41ft 6in)
Wingspan: 14.41m (47ft 3in)
Height: 4.29m (14ft 1in)
Weight: 9800kg (21,605lb) maximum take-off
Powerplant: 2 × Junkers Jumo 004B-1 axial flow turbojet engines, 8.83kN (1990lbf) thrust each
Maximum speed: 735km/h (456mph)
Range: 1556km (967 miles) with 500kg (1100lb) bomb load
Service ceiling: 10,000m (33,000ft)
Armament: 3 x 500kg (1100lb) SC 500J bombs

Ar 234 B-2
This Arado Ar 234 B-2 jet bomber bears the small black formation identification markings 'F1' markings next to the German Cross that denotes Kampfgeschwader 76.

Alt Lönnewitz in Saxony. The facility delivered all its contracted 200 airplanes by the end of December 1944.

TRANSPORT AND RECONNAISSANCE AIRCRAFT

A large proportion of the contribution that German air power made to the war effort was undertaken by combat aircraft – fighters, ground-attack platforms and bombers. However, the often unsung duties undertaken by German transport and reconnaissance aircraft also played an important part.

Aerial transport played a key role in augmenting ground resupply, getting critical stores quickly to locations where these supplies were desperately needed, such as encircled Stalingrad. Strategic, operational and tactical reconnaissance helped facilitate the successful conduct of the operations carried out by all three branches of the German Armed Forces.

Junkers Ju 52 transport aircraft

The Ju 52 was a tri-motor aircraft with low-set cantilever wings originally developed for civilian use. Militarily, it was employed primarily as a transport plane. The airframe sported one engine in its nose and one each mounted in the wing leading edges. The standard Ju 52/3m was powered by three 574kW (770hp) BMW 132 engines. During the 1930s, 200 civilian Ju 52s were used as 17-seater airliners or small cargo planes. In 1935–44, moreover, Junkers produced 3305 militarized Ju 52s, insufficient to meet the armed forces' aerial transportation demands. The Ju 52/3m transporter could carry 12 stretchers when operating as a medical evacuation platform.

Alternatively, the Ju 52 could carry two nine-man infantry sections, so two aircraft could move a complete platoon with its headquarters squad. The Ju 52 thus became the delivery platform for German paratrooper formations. In 1940, 500 of the 700 Ju 52s in service delivered German paratroopers to key strategic locations during the April 1940 and May–June 1940 German invasions of Norway and the West, respectively; in the May 1941 German airborne invasion of Crete, 350 Ju 52s delivered 12,000 paratroopers, but 150 of these aircraft were lost in the process.

The Germans also produced specialized variants of the standard Ju 52 3m transport. Some 70 Ju 52s seaplane variants, for example, were produced. Another such platform was the Ju 52/3m-MS 'Mine Searching' aircraft (and its similar seaplane version); these incorporated a 14m (46ft)-diameter degaussing ring, with which they detonated magnetic naval mines.

Ju 52s in flight

A formation of Junkers Ju 52 transport planes fly across the Mediterranean, possibly from Italy or Sicily to North Africa; by 1943 these reliable transport work-horses had become highly vulnerable to enemy fighter attack, and so required the protection provided by German interceptor aircraft.

Ju 52 production figures, 1939–44	
Year	**Number produced**
1939	145
1940	388
1941	502
1942	503
1943	887
1944	379

Around 350 Ju 52 transports – out of the 800 then in German service – were used in the winter 1942–43 airlift to resupply the German Sixth Army encircled in Stalingrad; during this operation, some 216 Ju 52s were shot down by enemy fighters or ground-based flak guns. Subsequently, in 1942–43, Ju 52s made up to 150 sorties a day transporting supplies from Italy to the Axis forces in North Africa. During April 1943, a sustained Allied counter-air onslaught destroyed 132 Ju 52s in the air or on the ground in southern Italy.

The last major Ju 52 air assault of the war occurred in the December 1944 Battle of the Bulge; 122 Ju 52s, many piloted by inexperienced crew, dropped some 1100 paratroopers in a widely dispersed fashion, with only 300 arriving at the intended objective.

Junkers Ju 52/3m g7e
Type: Transport
Crew: 3 (2 pilots, 1 radio operator)
Length: 18.9m (62ft)
Wingspan: 29.25m (95ft 10in)
Height: 4.5m (14ft 10in)
Weight: 9200kg (20,270lb) loaded
Powerplant: 3 × 533kW (715hp) BMW 132T radial engines
Maximum speed: 265km/h (165mph) at sea level
Range: 870km (540 miles)
Service ceiling: 5490m (18,000ft)
Armament: 1 × 13mm (0.5in) MG 131 machine gun in a dorsal position

Ju 52/3m
This Ju 52 belonged to the 2nd Squadron of Kampfgeschwader zbV 1 ('Special Purpose Bomber Wing' 1) and was based at the Greek island of Milos in the western Aegean during the spring 1941 Axis airborne invasion of Crete.

Focke Wulf FW 200 Condor reconnaissance/maritime/ transport aircraft

The Condor was an all-metal, four-engined, long-range monoplane. It performed three main long-range roles: as a reconnaissance aircraft, an anti-shipping/maritime patrol bomber, and as a transport aeroplane. The development of the Condor originated in a long-range, high-altitude civil airliner that could fly passengers from Germany to the United States. This civil Condor undertook trans-Atlantic flights in 1938–39 carrying 24 passengers.

After the outbreak of war, Focke Wulf produced slightly modified militarized Fw 200 C-1 Condors. These aircraft were powered by four potent 895kW (1200hp) BMW Bramo 323 nine-cylinder air-cooled engines. With this propulsion, the Condor could obtain a maximum speed of 360km/h (224mph) and an impressive maximum cruising range of 3560km (2212 miles).

In addition to the C-1, the Germans produced small numbers of the C-2 to C-6 sub-variants. The precise armament mounted in the aircraft differed between the sub-variants, but typically a Condor mounted between three and six 20mm (0.79in) MG 151/20 cannon and 7.92mm (0.31in) MG 15 or 13mm (0.5in) MG131 machine guns; these were located in various positions on the main fuselage and the latter's under-slung weapons-pod nacelle.

In total, Focke Wulf completed 276 Condors during the period 1939–44. In terms of the operational use of these aircraft, in 1940–42, the Fw 200 carried out long-range sweeps of the Atlantic using their radar to search for Allied convoys that they could then direct

Fw 200 C-6

A few Fw 200 C-series aircraft were converted to carry one Henschel Hs 293 radio-controlled winged anti-shipping glide-bomb under each of their outer engine nacelles; the antenna for the FuG 203 Kehl missile control transmitter was fitted to this variant's nose cone.

Focke-Wulf Fw 200 C-6

Type: Maritime reconnaissance
Crew: 5
Length: 23.45m (76ft 11in)
Wingspan: 32.85m (107ft 9in)
Height: 6.3m (20ft 8in)
Weight: 22,714kg (50,057lb) maximum take-off
Powerplant: 4 × 895kW (1200hp) BMW/Bramo 323R-2 nine-cylinder single-row air-cooled radial engine
Maximum speed: 360km/h (224mph)
Range: 3560km (2212 miles)
Service ceiling: 6000m (19,700ft)
Armament: 1 × 20mm (0.79in) MG 151/20 cannon in forward gondola; 2 × 13mm (0.5in) MG 131 machine gun dorsal turret; up to 5400kg (11,905lb) of bombs

Me 323 D-6
The large nose cone of the monstrous Me 323 opened up to permit access to the large forward-located cargo bay. The D-6 variant featured variable-pitch, three-blade Ratier propellers, as opposed to the two-bladed fixed-pitch Heine propellers of the D-1.

Messerschmitt Me 323 D-6
Type: Transport
Crew: 5, plus 130 troops or 10–12 tonnes (9.8–11.8 tons) of equipment
Length: 28.2m (92ft 4in)
Wingspan: 55.2m (181ft)
Height: 10.15m (33ft 3in)
Weight: 29,500kg (65,000lb) maximum take-off
Powerplant: 6 × 868kW (1164hp) Gnome-Rhône 14N-48/49, 1180 PS for take-off
Maximum speed: 219km/h (136mph)
Range: 800km (500 miles)
Service ceiling: 4000m (13,123ft)
Armament: 5 x 7.92mm (0.31in) MG 15 or 13mm (0.5in) MG 131 MGs

U-boats to attack; they could also drop a small payload of bombs or naval mines.

As Allied counter-threats magnified from 1942 onwards, Condors were increasingly redirected to carry out transport roles in support of ground forces. Significant numbers of Fw 200s, for example, were employed during the airlift undertaken in winter 1942–43 to supply the German Sixth Army, surrounded in Stalingrad. After the loss of the French Atlantic coast ports in 1944, naval strategic reconnaissance dwindled considerably; the Condor ceased to be used in this role as well. A single Condor was used as Hitler's personal transport plane, while six Model C-8s were converted to be the launch and guidance platform for the Henschel Hs 293 guided anti-ship missile.

Messerschmitt Me 323 Gigant ('Giant') transport aircraft

The aptly named Messerschmitt Me 323 Gigant ('Giant') was the largest military aircraft to serve during the 1939–45 war. This extremely large transport could carry a 12-tonne (11.8-ton) payload, sufficient to carry a 15cm (5.9in) sFH 18 artillery piece, an 8.8cm (3.4in) Flak gun, or 130 personnel. The Gigant was developed from the Me 321 military glider. The latter was developed in 1940–41 and was used for transportation purposes during the 1941 Axis invasion of the Soviet Union.

The Me 323 D production aircraft featured massive high-mounted semi-cantilever wings in the middle of its large rectangular fuselage. To keep weight down, much of the aircraft was made from plywood, metal poles and canvas. The aircraft was powered by six Gnome-Rhône 14N-48/49 14-cylinder air-cooled radial engines, with three mounted in each wing's leading edge. To even be able to take off successfully with a full payload, however, the aircraft needed the additional lift created by firing the four Walter HWK 109–500 rockets mounted on the wings.

The Model D mounted five 7.92mm (0.31in) MG 15 or 13mm (0.5in) MG 131 machine guns, two located in ball-mounts on the rear wing and three in the fuselage. The Gigant could fly at only a modest speed of 219km/h (136mph) at sea level, and less than this at altitude. The aircraft had a five-man crew, although two gunners could also be carried. The subsequent Me 323 D-2, D-6, E-1 and E-2 variants differed merely in featuring modified engines or propeller blades.

A total of 198 Gigant aeroplanes were built from scratch, plus another 30 were converted from existing Me 321 gliders. On 22 April 1943, a group of 27 fully loaded Gigant aircraft were intercepted by Allied fighters flying over the Sicilian Straights and 21 were shot down.

Fieseler Fi 156 Storch ('Stork') reconnaissance aircraft

The Fieseler Fi 156 Storch was a small, high-winged, single-engined monoplane, intended for army liaison and tactical reconnaissance roles. The aircraft had an impressive Short Take-Off and Landing (STOL) capability; it needed just 25m (82ft) of space in which to land and just 50m (164ft) from which to take off. The wings of the Fi 156 also folded down, enabling the aircraft to be transported in a standard military trailer.

The two-crew Fi 156 was powered by a single nose-mounted 179kW (240hp) Argus As 10 inverted-V8 engine. With this, the aircraft could achieve a modest top speed of 175km/h (109mph) and a maximum operational range of 380km (240 miles). Its renowned ability to 'loiter' over the battlefield was facilitated by its incredibly low stalling speed of just 50km/h (31mph). The Storch was armed

Air lift

Image captured during the North African campaign, with a landed Fieseler Fi-156 (right), from which a casualty has been evacuated to a nearby Junkers Ju 52 (left). Both planes clearly display the Red Cross insignia on their fuselages.

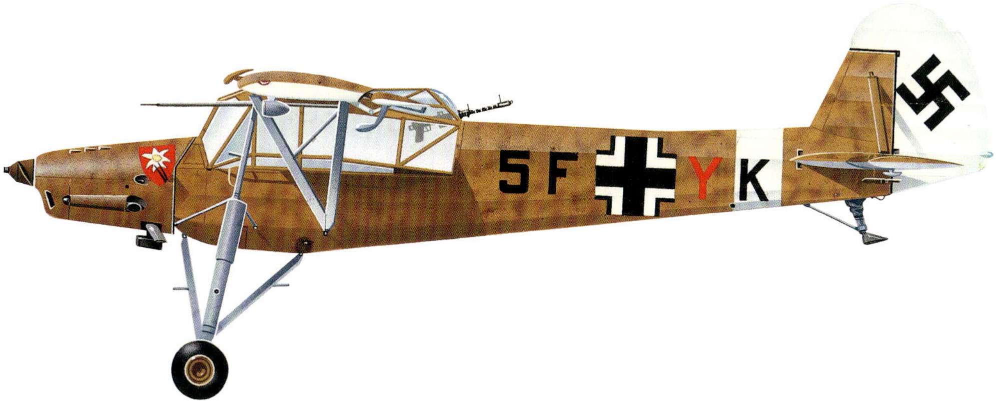

Fi 156, tropical colour scheme
This Fieseler Fi 156 C-3/Tropisch ('Tropical') variant serving with the Afrika Korps has its squadron code letter 'K' painted in black upon the white vertical theatre-recognition band applied to the overall sand-coloured camouflage scheme.

Fieseler Fi 156
Type: Reconnaissance aircraft
Crew: 2
Length: 9.9m (32ft 6in)
Wingspan: 14.3m (46ft 9in)
Height: 3.1m (10ft)
Weight: 1260kg (2780lb) loaded
Powerplant: 1 × 180kW (240hp) Argus As 10 air-cooled inverted V8 engine
Maximum speed: 175km/h (109mph)
Range: 380km (240 miles)
Service ceiling: 4600m (15,000ft)
Armament: 1 × 7.92mm (0.31in) MG 15 machine gun

with a solitary MG 15 machine gun, operated by the rear-facing gunner who was located immediately behind the pilot in the large cockpit nacelle.

Between 1937 and 1942, production factories in Kassel, Budweis (in Bohemia) and Puteaux in France between them manufactured 2905 Storch aircraft. Some 1230 examples of this production run were the principal variant, the C-3; some 286 C-1 and 239 C-2 models were also produced. The Storch was used by higher ground commanders as a personal reconnaissance aircraft. This versatile platform could also be used for many other tactical missions, such as air ambulance duties or for the insertion of agents behind enemy lines. The Storch, for example, famously participated in the daring September 1943 liberation of former Italian dictator Benito Mussolini from his Gran Sasso prison.

Henschel Hs 126 reconnaissance aircraft

A two-seat reconnaissance and observation monoplane, the Henschel Hs 126 entered Luftwaffe service in 1938; it remained in production until 1941. The aircraft was a development of the earlier Hs 122A recce platform. The Hs 126 sported a high parasol wing, under which sat the pilot in a protected cockpit with the rear-facing observer/gunner behind him in an open position. Powered by a Bramo 323 nine-cylinder radial engine, the aircraft could achieve a maximum speed at an altitude of some 356km/h (221mph). The Hs 126 also had an impressive STOL capability. By way of armament, the aircraft had a forward-firing 7.92mm (0.31in) MG 17, operated by the pilot, and a flexible 7.92mm (0.31in) MG 15, manned by the gunner; it could also carry a small

Hs 126 B-1
Finished in a white-washed winter camouflage scheme, this Hs 126 B-1 (platform 'P2+JZ') served with the 2nd (Army Cooperation) Squadron, Aufklärungsgruppe ('Reconnaissance Group') 21 on the Don River front, southern Eastern Front, during January 1943 on recce and glider towing duties.

Henschel Hs 126 B-1
Type: Reconnaissance aircraft
Crew: 2
Length: 10.9m (35ft 7in)
Wingspan: 14.5m (47ft 7in)
Height: 3.8m (12ft 4in)
Weight: 2030kg (4480lb) maximum take-off
Powerplant: 1 x 625kW (850hp) Bramo 323 9-cylinder radial engine
Maximum speed: 356km/h (221mph) at 3000m (9850ft)
Range: 998km (620 miles)
Service ceiling: 8530m (28,000ft)
Armament: 1 × forward-firing 7.92mm (0.31in) MG 17 machine gun; 1 × flexible 7.92mm (0.31in) MG 15 machine gun in the observer/gunner; up to 150kg (330lb) of bombs

bomb payload of some 150kg (330lb). The Hs 126 first saw operational service with the Condor Legion during the Spanish Civil War. It was then widely employed in the 1939 Polish and 1940 Western campaigns. By 1942, however, the Hs 126 had been largely superseded by the more effective Fi 156 Storch and Fw 189 Uhu aircraft. Thereafter, the Hs 126 was increasingly used as a glider-towing platform.

One of the aircraft's best-known missions occurred on 12 September 1943, when a group of 10 Hs 126s each towed a DFS 230 glider to the Gran Sasso massif in Italy. The gliders inserted a force of 103 German SS commandos that liberated Benito Mussolini from incarceration; the dictator was then flown to safety in a Fieseler Storch.

Focke Wulf FW 189 Uhu ('Owl') reconnaissance aircraft

The Focke Wulf Fw Uhu ('Owl') featured twin longitudinal booms extending from the wings on either side of the centre line, and a small central gondola-like fuselage with large 360-degree-view crew cockpits. The prototype of this three-seat short-range tactical reconnaissance aircraft first flew in 1938. The main production variant, the Fw 189 A-1, entered Luftwaffe service in 1940. Between then and 1944, three factories in Bremen, Bordeaux and Prague manufactured some 864 Uhu aircraft between them. The aircraft could be fitted with cameras; given its army cooperation and reconnaissance roles, it was soon nicknamed 'The Flying Eyes' by German Army personnel.

Fock Wulf Fw 189 A-1
Type: Reconnaissance aircraft
Length: 12m (39ft 4in)
Wingspan: 18.4m (60ft 4in)
Height: 3.7m (12ft)
Weight: 3950kg (8708lb)
Powerplant: 2 × 342kW (459hp) Argus As 410
Maximum speed: 357km/h (222mph) at 2600m (8530ft)
Range: 670km (416 miles)
Service ceiling: 8400m (27,550ft)
Armament: 2 × 7.92mm (0.31in) MG 17 machine guns mounted in the wing roots; 1 × 7.92mm (0.31in) MG 15 machine gun in dorsal flexible mount position

189 A-1
Two Fw 189 A-1s: (Top): aircraft 'V7+1E' of Aufklärungsgruppe 32 on the Eastern Front during early 1943; (bottom): as identified by the 'red devil' insignia on the engine, this aircraft served with the 1st Squadron, Reconnaissance Group 11, Ukraine 1943.

The initial main production model was designated the Fw 189 A-1. This variant was powered by two 342kW (459hp) Argus As 410 air-cooled inverted V-12 engines, one located in the front nose of each boom. By way of armament, the Model A-1 mounted two drum-fed flexible 7.92 (0.31in) MG 15 and two wing root-mounted MG 17 machine guns; it could also carry a 50kg (110lb) bomb payload.

Small numbers of seven slightly modified sub-variants were produced. These included 22 Fw 189 A-4 light ground-attack variants (with two twin 20mm/0.79in MG151/20 cannon), and 10 examples of the Fw 189 B-1 five-seat training version. Produced from 1942 onwards, the Fw 189 Model A-2 mounted two belt-fed twin-barrelled 7.92mm (0.31in) MG 81Z machine guns instead of the MG 15s.

Admiral Graf

CHAPTER 5

Naval Weapons

The Kriegsmarine – the German Navy – played a crucial role during World War II. Its key surface fleet assets – battleships, battlecruisers, pocket battleships and cruisers – spearheaded the above-the-waves 'surface raider' strategy of interdicting Allied supplies being transported from across the world to Britain or Russia. At the same time, German U-boat submarines waged a similar bitter and protracted interdiction offensive below the waves.

In the following section, we examine the seven key German capital ships: the battleships *Bismarck* and *Tirpitz*; the battlecruisers *Scharnhorst* and *Gneisenau*; and the pocket battleships *Deutschland* (*Lützow*), *Admiral Scheer* and *Admiral Graf Spee*.

Battleship *Bismarck*

Bismarck was the first vessel in the two-ship Bismarck class of battleships, the most potent surface vessels ever completed for the Kriegsmarine. The shipbuilding firm of Blohm & Voss in Hamburg laid down the keel of the ship in 1936. After feverish construction, she was launched in February 1939. During consequent fitting-out work, the ship's straight bow was replaced with a raked 'Atlantic' bow.

Bismarck was commissioned in August 1940 and spent four months on sea trials

Opposite:
***Admiral Graf Spee* launches**
On 30 June 1934, *Admiral Graf Spee*, the final ship in the Deutschland class of 'pocket battleships', was launched in front of a large crowd, the vast majority of whom are giving the newly-introduced raised-arm Nazi salute.

in the Baltic Sea. This mighty battleship's standard displacement was 42,369 tonnes (41,700 tons) unloaded. *Bismarck* was 251m (823ft) long, with a beam of 36m (118ft). Serviced by a crew of 2065, *Bismarck* was powered by three Blohm and Voss 36,815kW (49,370hp) steam turbines and could reach a maximum speed of 30 knots (55km/h; 34mph). The battleship sported eight 38cm (15in) SK C/34 guns in four paired turrets, two forward and two aft, as well as a secondary armament of 12 15cm (5.9in) SK C/28 guns. The battleship's hull was protected by armoured belts that were up to 320mm (12.6in) thick, while her turret fronts consisted of armour 360mm (14.2in) thick. The vessel also carried a catapult to launch her four Arado Ar-196 reconnaissance seaplanes.

During 19–23 May 1941, *Bismarck*, along with the heavy cruiser *Prinz Eugen*, left the Baltic and sailed through Norwegian waters, and thence between Iceland and Greenland through the Denmark Straits, where they were detected by HMS *Suffolk*. The 'surface raider' mission given to *Bismarck* was to get into the North Atlantic and sink as much Allied shipping as possible. On 24 May, the battlecruiser HMS *Hood* and battleship HMS *Prince of Wales* engaged the two German vessels; during the action, *Hood* was sunk and the damaged *Prince of Wales* was forced to retire. *Bismarck*, however, had also sustained damage during the engagement, including flooded compartments and leaking fuel tanks. Admiral Lütjens decided the ship would head towards a German-occupied French Atlantic port for repairs while *Prinz Eugen* continued with the raiding mission.

On 25 May, British carrier aircraft successfully engaged *Bismarck* as she headed southeast towards the Bay of Biscay at a reduced speed of 27 knots (50km/h; 31mph). These attacks severely damaged the vessel's port rudder and steering mechanisms, causing her to

Hitler salutes *Bismarck*

The German Führer, Adolf Hitler (centre), gives the Nazi salute to *Bismarck*; some of the ship's secondary and tertiary armament is visible adjacent to his out-stretched arm.

Bismarck

Displacement: 42,369 tonnes (41,700 tons); 50,300 tonnes (49,500 tons) full load
Dimensions: 251 x 36 x 9.3m (823 x 118 x 31ft)
Propulsion: triple shafts, 12 Wagner HP boilers, three Blohm & Voss geared turbines 110,450 kW (148,116hp)
Armament: 8 x 381mm (15in), 12 x 150mm (5.9in), 16 x 105mm (4.13in) guns; 16 x 37mm (1.46in), 12 x 20mm (0.79in) AA guns
Armour: belt: 80–320mm (3.1–12.6in); bulkheads: 45–220mm (1.7–8.6in); turrets: 180–360mm (7–14.2in), deck 80–120mm (1.7–4.3in)
Speed: 30 knots (55km/h; 34mph)
Aircraft carried: 4 × Arado Ar 196 floatplanes
Crew: 2065

Bismarck
A profile of *Bismarck* showing the tops of her main and secondary gun turrets painted yellow; according to one of the few survivors of the sinking this was done while the battleship was at sea during her fateful sortie into the Atlantic.

circle gently to port. After dawn on 27 May, two British battleships, two heavy cruisers and five destroyers moved in for the kill against the now stricken and non-manoeuvrable battleship. For two hours, an intense fury of close-range gunfire raged, with *Bismarck* being hit more than 400 times.

Eventually the cruiser *Dorsetshire* moved in for the fatal blow, mounting several successful close-range torpedo attacks. Subsequently, the wrecked and listing *Bismarck* heeled to port and capsized, probably after eventually being scuttled. Only 114 German personnel survived the sinking.

Battleship *Tirpitz*

Tirpitz was the second vessel in the Bismarck class of battleships to be completed. She was laid down at Wilhelmshaven in November 1936; the completed hull was launched in May 1939. She was commissioned in February 1941 and undertook sea trials in the Baltic.

Tirpitz mounted eight 38cm (15in) SK C/34 guns in two forward and two aft main twin turrets; her secondary armament consisted of 12 15cm (5.9in) SK C/28 guns. Thanks to the incorporation of some minor design modifications, *Tirpitz* had a standard

The wreck of the *Bismarck*

Discovery date	Location	Depth	Orientation
8 June 1989	650km (400 miles) west of Brest	4791m (15,719ft)	Upright

displacement of 43,588 tonnes (42,900 tons) – a little heavier than *Bismarck*. Her three Brown, Boveri & Cie geared steam turbines between them produced 119,905kW (160,796hp), more than those on the *Bismarck*, which enabled her to reach a slightly higher maximum cruising speed of 30.8 knots (57km/h; 35mph). During the war, *Tirpitz* went through several programmes of modification. Her original 12 2cm (0.79in) Flak guns, for example, were eventually augmented by a further 46 such weapons; she also incorporated eight above-water torpedo tubes.

In January 1942, *Tirpitz* moved to Trondheim in Norway, from where she could sally forth into the North Atlantic to devastate Allied shipping convoys. *Tirpitz* was berthed next to a cliff in the Faettenfjord with numerous protective underwater nets and anti-aircraft batteries deployed around her. Despite several planned sorties into the Denmark Straits, circumstances and lack of fuel worked to thwart these plans save for two sorties in March and June 1942. Next, in March 1943, *Tirpitz* and *Scharnhorst* bombarded Allied meteorological installations on the island of Spitzbergen.

In September 1943, eight British X-craft midget submarines attacked *Tirpitz* in the Faettenfjord. Two of the mini-submarines, X6 and X7, managed to place their mines under the battleship and detonate them. *Tirpitz* suffered significant damage, which took seven

Tirpitz

Profile of the battleship *Tirpitz*, sporting a dark and light grey geometrical camouflage scheme; five vents fed the steam generated by her 12 superheated boilers up through the funnel stack.

Tirpitz

Displacement: 43,588 tonnes (42,900 tons); 52,600 tonnes (51,800 tons) full load

Dimensions: 251 x 36 x 9.3m (823ft 6in x 118ft 1in x 30ft 6in)

Propulsion: triple shafts, three Brown-Boveri geared turbines, 12 Wagner superheated boilers, 119,905kW (160,796hp)

Armament: 8 x 380mm (15in), 12 x 150mm (5.9in), 16 x 105mm (4.13in) guns; 16 x 37mm (1.46in), 12 x 20mm (0.79in) AA guns

Armour: belt: 80–320mm (3.1–12.6in); bulkheads: 45–220mm (1.7–8.6in); barbettes: 220–340mm (8.6–13.4in); turrets: 180–360mm (7–14in); deck: 50–120mm (2–4.7in)

Speed: 30.8 knots (57km/h; 35mph)

Aircraft carried: 4 × Arado Ar 196 floatplanes

Crew: 2065

Armour protection on the *Tirpitz*				
Description	**Upper deck**	**Main deck**	**Main belt**	**Turret front**
Thickness	50mm (2in)	100–120mm (3.9–4.7in)	320mm (13in)	360mm (14in)

months to repair fully. On the very day in April 1944 on which the repaired *Tirpitz* set off for sea trials, 40 British dive-bombers attacked the battleship, inflicting yet more significant damage, including the destruction of two of her four main turrets.

Yet more air attacks ensued in mid-1944, which caused additional damage. Then, on 15 September, an RAF Lancaster bomber hit *Tirpitz* with a Tallboy bomb. One month later, the damaged battleship limped at 8 knots (15km/h; 9mph) to a new protected berth in the Tromsofjord. Here, *Tirpitz* was kept operational only as a powerful floating artillery battery with a reduced crew. On 12 November 1944, she was attacked by 32 Lancaster bombers that scored two direct hits. *Tirpitz* slowly started listing to port and eventually rolled over and buried her superstructure in the fjord bottom; some 1000 crew perished in the process.

Battlecruiser *Scharnhorst*

Scharnhorst was the first ship in the class of two battlecruisers of the same name. Laid down in 1935, the vessel was built by the Reichsmarinewerft of Wilhelmshaven and was launched in late 1936. In February 1939, she was officially commissioned into the Kriegsmarine and began her Baltic Sea operational trials. After this, her straight bow was replaced with a raked 'Atlantic' bow to improve her seaworthiness in rough waters.

Scharnhorst displaced 32,615 tonnes (32,100 tons) unloaded and 38,711 tonnes (38,100 tons) when fully loaded. Her main armament consisted of nine 28cm (11in) SK C/34 quick-firing guns in three treble turrets, with two turrets located forward and one aft; these guns could fire seven times in 120 seconds. Her secondary armament consisted of 12 15cm (5.9in) SK C/33 guns, as well as 40 (later 46) anti-aircraft guns. Her three Brown, Boveri & Cie super-heated steam turbines generated

Crew inspection
In the left foreground Kapitän zur See (Captain at Sea) Otto Ciliax, commander of the *Scharnhorst*, undertakes an inspection of the battlecruiser's crew sometime during 1939.

a combined total of 111,717kW (149,815hp), enabling her to reach a top cruising speed of 31 knots (57km/h; 36mph). She was considered a poor sea boat, however, being bow-heavy and often 'awash' fore and aft in rough seas; her first forward main gun turret could not operate in rough seas.

On 21 November 1939, *Scharnhorst* carried out her first operational mission, conducting a sweep between the Faroe Islands and Iceland; during this sortie, she sank the British armed merchant cruiser HMS *Rawalpindi*. Next, the battlecruiser covered the German invasion forces that landed at Narvik and Trondheim during the April 1940 German invasion of Norway. Subsequently, on 8 June, *Scharnhorst* and *Gneisenau* engaged the British aircraft carrier HMS *Glorious*. The former hit the carrier at the extremely long range of 24,200m (79,400ft). Between them, the German vessels sank the carrier and her two escorting destroyers while suffering only modest damage themselves.

Next, during January 1941, *Scharnhorst* and *Gneisenau* sallied forth from Kiel and broke into the North Atlantic, sinking several merchant ships before returning to the port of Brest in France on 22 March 1941. Next, in February 1942, both battlecruisers, plus the heavy cruiser *Prinz Eugen*, made a daring dash through the English Channel to reach Wilhelmshaven in northwestern Germany.

In March 1943, *Scharnhorst* sailed to Narvik in Norway, ready for further forays into the North Atlantic. Eventually, in late December 1943, *Scharnhorst* sailed to intercept a convoy bound for the Soviet Union. In what became known as the Battle of the North Cape, the vessel was engaged by

Reconnaissance aircraft

A crane loads an Ar-136 Arado floatplane, which had just landed in the adjacent sea, onto *Scharnhorst*. Some 526 such floatplanes served with the Kriegsmarine during World War II.

Scharnhorst

Displacement: 32,615 tonnes (32,100 tons); 38,711 tonnes (38,100 tons) full load
Dimensions: 235 x 30 x 9.7m (772ft x 98ft 5in x 31ft 9in)
Propulsion: triple shafts, 12 Wagner HP boilers, three Brown-Boveri geared turbines, 111,717kW (149,815hp)
Armament: 9 x 280mm (11in), 12 x 150mm (5.9in), 14 x 105mm (4.13in) guns; 16 x 37mm (1.46in) and 10 x 20mm (0.79in) AA guns; 6 x 533mm (21in) torpedo tubes
Armour: belt: 200–350mm (7.9–13.8in); bulkheads: 150–200mm (5.9–7.9in); barbettes: 200–350mm (7.9–13.8in); turrets: 200–350mm (7.9–13.8in); deck: 20–50mm (0.79–2in)
Speed: 31 knots (57km/h; 36mph)
Aircraft carried: 3 x Arado Ar 196A
Crew: 1986

Scharnhorst
A profile of *Scharnhorst* showing her steam boilers (in brown) and her BBC turbines (in green), as well as the heavily-armoured magazine for her second forward main gun turret, (designated 'Bruno').

three British cruisers and the battleship *Duke of York*. Crippled by dozens of British hits and at least 18 torpedo strikes, the stricken *Scharnhorst* capsized to starboard and sank. In freezing northern waters, only 36 of the vessel's crew of 1986 survived this encounter.

Battlecruiser *Gneisenau*

The German battlecruiser *Gneisenau* was laid down in February 1934 at the Deutsche Werke shipyard at Kiel; she was the final ship in the *Scharnhorst* class of two vessels. She was launched in December 1936 and was commissioned into the German Navy in May 1938.

Gneisenau's completion was delayed due to the incorporation of significant design modifications during 1937. These changes meant her standard displacement was 32,310 tonnes (31,800 tons), somewhat lighter than her sister ship *Scharnhorst*. *Gneisenau* mounted the same arrangement of three treble turrets with 28cm (11in) SK C/34 guns as seen in her sister ship. The vessel also differed from her sister ship in that she had three Germania super-heated steam boilers that between them produced 121,400kW (162,800hp). Her top speed was 31.1 knots (58km/h; 36mph); steaming at a sustained steady speed of 19 knots (35km/h; 22mph), *Gneisenau* could sail 6200 nautical miles (11,480km; 7135 miles) on her fuel load – slightly less than her sister ship.

After completion of her sea trials, the vessel undertook her first operational sortie; in October 1939, she sallied forth into the Atlantic as part of a squadron of

Gneisenau

This photograph of *Gneisenau* ably depicts the large size, relatively speaking, of her treble 28cm (11in) SK C/34 gun turrets, as well as the curves of her 'Atlantic Bow', designed to improve sea-worthiness.

Gneisenau

Displacement: 32,615 tonnes (32,100 tons); 38,711 tonnes (38,100 tons) full load

Dimensions: 235 x 30 x 9.7m (772ft x 98ft 5in x 31ft 9in)

Propulsion: triple shafts, Germania geared turbines with single reduction three-bladed propellers, 121,400kW (162,800hp)

Armament: 9 x 280mm (11in), 12 x 150mm (5.9in), 14 x 105mm (4.13in) guns; 16 x 37mm (1.46in) and 10 x 20mm (0.79in) AA guns; 6 x 533mm (21in) torpedo tubes

Armour: belt: 200–350mm (7.9–13.8in); bulkheads: 150–200mm (5.9–7.9in); barbettes: 200–350mm (7.9–13.8in); turrets: 200–350mm (7.9–13.8in); deck: 20–50mm (0.79–2in)

Speed: 31 knots (57km/h; 36mph)

Aircraft carried: 3 x Arado Ar 196A

Crew: 1669

13 German surface vessels. Subsequently, during June 1940, *Gneisenau*, along with her sister ship *Scharnhorst*, engaged and sank the British aircraft carrier HMS *Glorious*; in the process, however, *Gneisenau* suffered minor damage herself. Next, the two battlecruisers sailed from Kiel into the North Atlantic during January–March 1941; after a successful episode of 'surface raiding', *Gneisenau* and *Scharnhorst* returned to Brest for repairs.

Subsequently, during February 1942, *Gneisenau*, plus her sister ship and the heavy cruiser *Prinz Eugen*, audaciously sailed through the English Channel to reach the port of Wilhelmshaven. *Gneisenau* went into dry dock at Kiel so that significant repairs could be undertaken.

However, in late 1942, an Allied aerial bombing strike on Kiel harbour inflicted serious damage on *Gneisenau*. The repairs required were so extensive the Germans decided to refit the vessel with 38cm (15in) gun turrets and her original 28cm (11in) gun turrets were removed.

***Gneisenau* gunnery ranges**			
Armament type	28cm (11in) SK C/34	15cm (5.9in) SK C/28	10.5cm (4.1in) SK C/33
Maximum range of armament	42,600m (139,760ft)	22,000m (72,180ft)	17,700m (58,070ft)

In April 1942, the non-operational *Gneisenau* sailed to Gotenhafen in East Prussia for extensive repairs. In late 1943, however, after the sinking of *Scharnhorst*, Hitler ordered repair work on the vessel suspended, and she languished in Gotenhafen harbour.

Finally, during early 1945, the Germans used the non-operational vessel as a floating artillery battery as the advancing Soviet ground forces closed in on the port. Finally, on 27 March 1945, the Germans scuttled the battlecruiser in the harbour to prevent Soviet naval assets from using the facilities once the port had been captured.

Pocket Battleship *Deutschland* (*Lützow*)

Deutschland was the first of three vessels in the *Deutschland* class of 'armoured ships' or 'pocket battleships' that (unusually for the time) were powered by diesel engines; in 1940, however, the Germans reclassified them as heavy cruisers. They were constructed with some pretence at obeying the restrictions imposed on Kriegsmarine surface ship construction by the Treaty of Versailles.

Officially, the standard displacement of *Deutschland* was declared at 10,161 tonnes (10,000 tons) to comply with treaty restrictions; in reality, however,

Deutschland
Displacement: 10,770 tonnes (10,600 tons); 14,520 tonnes (14,290 tons) full load
Dimensions: 186 x 21.7 x 7.25m (610ft x 71ft x 24ft)
Propulsion: three turbines, 101,320kW (136,000hp)
Armament: 8 x 203mm (8in); 12 x 105mm (4.1in); 45 x AA guns; 12 x 533mm (21in) torpedo
Armour: belt: 60–80mm (2.4–3.1in); bulkheads: 40–45mm (1.6–1.8in); turrets: 85–140mm (3.3–5.5in); conning tower: 50–150mm (2–5.9in); deck: 40–45mm (1.6–1.8in)
Speed: 33.5 knots (62km/h; 39mph)
Aircraft carried: 2 x Arado Ar 196 seaplanes
Crew: 619

Deutschland
This profile of *Deutschland* reveals the modest size of her superstructure and bridge, all part of the scheme to maximize firepower and protection while ostensibly keeping within the 10,000-tonne Treaty displacement limit.

her standard displacement was 10,770 tonnes (10,600 tons). The design of these vessels sought to maximize armament and armour protection while keeping down displacement; this was mainly achieved by welding the hull rather than using rivetting.

Deutschland was laid down at Kiel in 1929 and was launched in May 1931. The finished vessel commenced her sea trials in November 1932 and was officially commissioned into the Navy in April 1933. She mounted six 28cm (11in) SK C/28 guns in one fore and one aft treble turret; these weapons could obtain a maximum range of 36,475m (119,670ft). The vessel's main armoured belt was 80mm (3.1in) thick.

When war broke out in September 1939, *Deutschland* was at sea and had a reasonably successful surface raiding sortie before returning to Germany. She was then renamed *Lützow* after the heavy cruiser of the same name was sold to the Soviet Union.

Lützow participated in the April 1940 German invasion of Norway, but was torpedoed and badly damaged by a British submarine on her return to Germany. Repairs on the vessel were completed only in March 1941, after which she returned to Norway. En route, she was torpedoed by a British bomber and subsequently spent over a year in Germany being repaired. On her return to Norway, she participated in the sortie against Convoy PQ 17, but then ran aground and had to return to Germany for yet more repairs. After another return to Norway, she fought in the Battle of the Barents Sea with the heavy cruiser *Admiral Hipper*.

Thereafter, *Lützow* remained in the Baltic for the rest of the war. In early 1945, she provided fire support to the German ground forces that were desperately defending Prussia and Pomerania against the inexorable Soviet advance westwards. In April, she was deployed in the Kaiserfahrt canal near Swinemünde when she was badly damaged by RAF bombers. Finally, on 4 May 1945, as German forces retreated from the area, *Lützow* was scuttled by her crew.

Pocket Battleship *Admiral Scheer*

Admiral Scheer was the second ship in the three-vessel Deutschland class of pocket battleships, later redesignated as heavy cruisers. She was laid down at the Reichsmarinewerft in Wilhelmshaven during June 1931 and was commissioned in November 1934.

The vessel had a standard displacement of 13,660 tonnes (14,895 tons), nearly one-quarter heavier than her sister ship, *Deutschland*. This extra weight was partly because *Admiral Scheer* had a beam 65cm (26in) wider, but also because she featured enhanced armoured protection.

She undertook a number of patrols in Spanish waters during the 1936–39 Spanish Civil War. Subsequently, her anti-aircraft weaponry was significantly augmented. Next, during October 1940, she sortied out into the southern Atlantic Ocean.

Over the next five months, and after rendezvousing with several German supply vessels, *Admiral Scheer* sank 17 merchant vessels during a successful 'surface raiding' sortie that was an

Admiral Scheer
In this fine view of the forward-port side of the pocket-battleship *Admiral Scheer*, her two rows of hull portholes and twin anchors can clearly be seen.

Admiral Scheer
Displacement: 13,660 tonnes (14,895 tons); 16,154 tonnes (15,900 tons) full load
Dimensions: 186 x 20.6 x 7.2m (610ft 3in x 67ft 6in x 23ft 7in)
Propulsion: quadruple shafts, four 9-cylinder double-acting 2-stroke MAN diesel engines, Vulcan gearboxes, 40,268kW (54,000shp)
Armament: 6 x 54.5-calibre/280mm (11in), 8 x 150mm (5.9in), 3 x 88mm (3.5in) guns; 8 x 500mm (19.7in) torpedo tubes
Armour: belt: 60–80mm (2.4–3.1in); bulkheads: 40–45mm (1.6–1.8in); turrets: 85–140mm (3.3–5.5in); conning tower: 50–150mm (2–5.9in); deck 40–45mm (1.6–1.8in)
Speed: 28 knots (52km/h; 32mph)
Aircraft carried: 2 x Arado Ar 196 seaplanes
Crew: 619

extraordinary 46,418 nautical miles (85,966km; 53,417 miles) long.

Next, in February 1942, the vessel sailed from German waters to Narvik in Norway. While located here, the heavy cruiser participated in the abortive German interception mission mounted against the Allied convoy PQ 17. Thereafter, in early 1943, she returned to dry dock in Wilhelmshaven for major repairs. While in dry dock, the heavy cruiser was damaged by an Allied aerial bombing attack. To avoid further damage caused by aerial attacks, *Admiral Scheer* then sailed in a barely operational state into the Baltic. She berthed in Swinemünde harbour on Usedom Island, located in the estuary of the River Oder. Hitler now ordered the decommissioning of the German capital ships, and *Admiral Scheer* took on cadet-training activities.

In early 1945, the recommissioned vessel, with augmented Flak capability, provided much-needed fire support to the German ground forces defending the coastal regions of East Prussia and Pomerania. In addition, she

***Admiral Scheer* launches**

The pocket battleship *Admiral Scheer* slides slowly down the slipway into the still waters of Wilhelmshaven dockyard during her official launching on 1 April 1933.

participated in Operation 'Hannibal': the evacuation of wounded military personnel and German civilians from areas close to the Soviet advance back to safety in northern Germany. After returning to Kiel to have her worn-out main gun barrels exchanged, she was attacked by RAF bombers on 9 April 1945. Hit several times, *Admiral Scheer* slowly began to list to starboard until eventually she capsized.

Anti-aircraft capabilities, *Admiral Scheer*	
Year	**Anti-aircraft weapons**
1934	3 x 8.8cm (3.4in) L/45
1935	6 x 8.8cm (3.4in) L/78
1940	6 x 10.5cm (4.13in) C/33, 4 x twin 3.7cm (1.46in) C/30, 28 x 2cm (0.79in) Flak 20
1945	6 x 4cm (1.57in), 8 x 3.7cm (1.46in) C/30, 32 x 2cm (0.79in) Flak 30

Pocket Battleship *Admiral Graf Spee*

Admiral Graf Spee, the final ship in the Deutschland class, had a standard displacement of 14,890 tonnes (14,648 tons) – 45 per cent heavier than *Deutschland*. This extra weight was partly caused by the *Admiral Graf Spee* having a 100mm (3.9in)-thick main armoured belt and being 96cm (38in) wider in the beam.

The vessel was laid down at Wilhelmshaven's dockyard during October 1932 and launched on 30 June 1934. Fitting out continued and, on 6 January 1936, the *Admiral Graf Spee* was officially commissioned into the German Navy. The pocket battleship mounted the same main armament as her two sister vessels: six 28cm (11in) SK C/28 guns in one fore and one aft treble turret. She featured the heaviest levels of protection within the three-vessel class. Her turret fronts had 140mm (5.5in) armour, her main belt was 80–100mm (3.1–3.9in) thick, and her decking was protected by plates 17–45mm (0.67–1.8in) thick. *Admiral Graf Spee* had four sets of MAN double diesel engines that combined produced 38,813kW (52,050hp) of propulsive energy. This enabled a cruising speed of 28.5 knots (53km/h; 33mph).

During 1936–38, *Admiral Graf Spee* spent five periods of time patrolling off the Spanish coast to uphold the non-intervention agreement signed in August 1936 in relation to the ongoing Spanish Civil War. In August 1939, the vessel departed for the South Atlantic and was on station when war broke out. On 26 September, *Admiral*

Graf Spee sank her first merchant ship, the cargo vessel *Clement*. During October, some eight Allied naval task forces began the hunt for the pocket battleship, which in the meantime had captured and/or sunk five merchantmen. After a sortie into the Indian Ocean in mid-November, she returned to the South Atlantic.

Early on 13 December, *Admiral Graf Spee* encountered the British heavy cruiser HMS *Exeter*, together with the light cruisers *Ajax* and *Achilles*. In the ensuing two-hour Battle of the River Plate, the pocket battleship hit *Exeter* six times with its main guns, causing extensive damage, and *Ajax* twice. *Admiral Graf Spee* was hit 68 times during the engagement, suffering significant damage to her diesel oil purification and desalination plants. The ship put into neutral Montevideo harbour for short-term repairs. In the meantime, British intelligence leaked exaggerated news of a powerful British fleet being assembled off the estuary of the River Plate. Rather than face defeat in battle, on 18 December a skeleton crew sailed the vessel out into the estuary and detonated powerful explosive charges, scuttling the vessel in shallow waters.

Admiral Graf Spee
Displacement: 14,890 tonnes (14,648 tons); 16,020 tonnes (16,280 tons) full load
Dimensions: 186 x 20.6 x 7.2m (610ft 3in x 67ft 6in x 23ft 7in)
Propulsion: quadruple shafts, four 9-cylinder double-acting 2-stroke MAN diesel engines, Vulcan gearboxes, 38,813kW (52,050hp)
Armament: 6 x 54.5-calibre/280mm (11in), 8 x 150mm (5.9in), 3 x 88mm (3.5in) guns; 8 x 500mm (19.7in) torpedo tubes
Armour: belt: 80–100mm (3.1–3.9in); bulkheads: 40–45mm (1.6–1.8in); turrets: 85–140mm (3.3–5.5in); conning tower: 50–150mm (2–5.9in); deck: 40–45mm (1.6–1.8in)
Speed: 28.5 knots (53km/h; 33mph)
Aircraft carried: 2 x Arado Ar 196 seaplanes
Crew: 619

Admiral Graf Spee
In this peace-time view of *Admiral Graf Spee*, the vessel is bedecked with flags. Note the Arado Ar-196 float-plane stowed on a platform amidships; in wartime this conducted valuable reconnaissance missions for the vessel.

CRUISERS

Germany's 'surface raider' strategy to interdict merchant vessels bringing supplies to Britain and the Soviet Union was carried out primarily by Germany's capital ships – battleships, battlecruisers and pocket battleships. But Germany's fleet of cruisers also played a part in this strategy. Two of Germany's three operational heavy cruisers – *Admiral Hipper* and *Prinz Eugen* – participated in 'surface raiding' missions, either alone or in escort to one or more capital ships. The Kriegsmarine, in addition, possessed six light cruisers: *Emden, Köln, Karlsruhe, Königberg, Leipzig* and *Nürnberg*. These vessels lacked the firepower and operational range to typically undertake solo surface raiding missions, but they did sometimes escort Germany's capital ships and/or heavy cruisers on such missions.

Heavy Cruiser *Admiral Hipper*

Admiral Hipper was the first ship of five in the class named after her. The vessel was laid down in Hamburg in July 1935 and was launched in February 1937. After completing her fitting out, the ship was commissioned on 29 April 1939 and declared operational on 17 February 1940.

Hipper had a standard displacement of 16,170 tonnes (15,910 tons) and featured three steam turbines that enabled her to achieve a top cruising speed of 32 knots (59km/h; 37 mph). Her main armament was eight 20.3cm (8in) SK L/60 guns mounted in two forward and two aft super-firing twin turrets. In April 1940, *Hipper* escorted amphibious landing forces to Trondheim as part of the German invasion of Norway, during which she sank the British destroyer *Glowworm*. Next, in November–December 1940, *Hipper* broke out through the Denmark Straits into the Northern Atlantic, where she sank one merchantman before returning to the port of Brest.

During her February 1941 sortie into the North Atlantic, *Hipper* sank eight merchant vessels, before spending seven months in dry dock in Kiel. In March 1942, *Hipper* sailed to Trondheim in Norway. On 3 July 1942, *Hipper,* along with *Tirpitz* and the former pocket battleships *Lützow* and *Admiral Scheer*, unsuccessfully attempted to decimate the Allied convoy PQ 17.

In December 1942, *Hipper* and *Lützow* attacked convoy JW 51B and was engaged by two British cruisers in what became known as the Battle of the Barents Sea. In the wake of this unsuccessful encounter, Hitler ordered the decommissioning of Germany's remaining large surface assets. Decommissioned in the Baltic during 1943–44, in early 1945 *Hipper* sailed to Kiel for a major refit to make her operational again.

On 3 May, RAF bombers severely damaged her at Kiel and her crew scuttled the crippled vessel.

Heavy Cruiser *Prinz Eugen*

The third vessel in the Hipper class was *Prinz Eugen*. The ship was laid down in April 1936 at Kiel, was launched on 22 August 1938 and was commissioned on 1 August 1940. During 1940–41, *Prinz Eugen* undertook sea trials and crew

Admiral Hipper

A team of crewmen employ a long rod to clean the barrels of *Admiral Hipper*'s main guns, one of many laborious routine maintenance jobs required to keep the ship at optimum combat-readiness.

Admiral Hipper
Displacement: 16,170 tonnes (15,910 tons)
Dimensions: 202.8 x 21.3 x 7.2m (665ft 4in x 69ft 11in x 24ft)
Propulsion: 3 × Blohm & Voss steam turbines, 3 × three-blade propellers, 98MW (132,000shp)
Armament: 8 × 20.3cm (8in) guns, 12 × 10.5cm (4.13in) SK C/33 guns, 12 × 3.7cm (1.46in) SK C/30 guns, 8 × 20mm (0.79in) C/30 guns (20 × 1), 6 × 53.3cm (21in) torpedo tubes
Armour: belt: 70–80mm (2.8–3.1in); armour deck: 20–50mm (0.79–1.97in); turret faces: 105mm (4.13in)
Speed: 32 knots (59km/h; 37mph)
Aircraft carried: 3 x Arado Ar 196 seaplanes
Crew: 42 officers, 1340 enlisted men

training in the Baltic. Her first 'surface raider' mission commenced on 19–23 May 1941, when she accompanied *Bismarck* through Norwegian waters and thence the Denmark Straits, where they were engaged by HMS *Hood* and HMS *Prince of Wales*.

On 24 May, the damaged *Bismarck* headed southeast towards a German-occupied French Atlantic port for repairs while *Prinz Eugen* continued southward. While sailing south on 27 May, the day *Bismarck* was sunk, the cruiser's crew discovered significant engines problems, and the vessel turned eastward back toward Brest for repairs. Over the next five days, she eluded numerous Allied task groups searching for her and reached Brest safely, where she entered dry dock for repairs.

Prinz Eugen

This fantastic frontal image of the heavy cruiser *Prinz Eugen* ably illustrates the height of the front superstructure, the form of the front 'Anton' gun turret, and the elegant lines of the 'Atlantic Bow'.

Prinz Eugen

Displacement: 16,970 tonnes (16,700 tons)

Dimensions: 212.5 x 21.7 x 7.2m (697ft 2in x 71ft 2in x 24ft)

Propulsion: 3 × Blohm & Voss steam turbines, 3 × three-blade propellers, 101MW (135,000shp)

Armament: 8 × 20.3cm (8in) guns, 12 × 10.5cm (4.13in) SK C/33 guns, 12 × 3.7cm (1.46in) SK C/30 guns, 8 × 2cm (0.79in) C/30 guns (20 × 1), 12 × 53.3cm (21in) torpedo tubes

Armour: belt: 70–80mm (2.8–3.1in); armour deck: 20–50mm (0.79–1.97in); turret faces: 105mm (4.1in)

Speed: 33.4 knots (62km/h; 38mph)

Aircraft carried: 3 x Arado Ar 196 seaplanes

Crew: 42 officers, 1340 enlisted men

Subsequently, during February 1942, *Prinz Eugen*, *Gneisenau* and *Scharnhorst* audaciously sailed through the English Channel to reach Wilhelmshaven. Next, during much of 1943, the semi-decommissioned vessel undertook cadet-training missions in the Baltic Sea. During late 1943 and throughout 1944, the recommissioned heavy cruiser undertook fire support missions for the German Army fighting on the northern sector of the Eastern Front, as well as evacuating wounded German military personnel and civilians from areas threatened by the Soviet advance. In March 1945, moreover, she fired 2447 main gun rounds in support of the Army's defence of the Bay of Danzig. On 13–20 April 1945, *Prinz Eugen* sailed from Swinemünde to Copenhagen, where critical fuel shortages prevented her from sailing again. She capitulated on 5 May.

Heavy Cruiser *Seydlitz*

The fourth vessel in the Hipper heavy cruiser class was the *Seydlitz*. Like her sister ship *Lützow*, *Seydlitz* was designed as a more lightly armed variant of the Hipper class, but was then redesigned as identical to the rest of her four sister vessels. The heavy cruiser was laid down in the Deschimag dockyard in Bremen in December 1936. After a lengthy construction programme, the vessel was finally launched in January 1939. By mid-1940, *Seydlitz* was 95 per cent complete, but then further work on her was abandoned.

As designed, she sported eight 20.3cm (8in) SK C/33 L/60 guns, had a standard displacement of 17,600 tonnes (17,300 tons), and a top cruising speed of 32 knots (59km/h; 37mph). The almost complete vessel languished in Bremen's harbour until March 1942, when it was decided to convert her to an auxiliary aircraft carrier. Renamed *Weser*, the ship had been only partly converted when work was stopped in March 1943. The incomplete carrier was later transferred to *Königsberg*, where, on 29 January 1945, she was scuttled as the Soviet advance reached the harbour area.

Heavy Cruiser *Lützow*

Lützow was the fifth and final vessel in the Hipper class of heavy cruisers. Originally she was conceived as a more lightly armed variant of the class, but then redesigned as identical to the rest of her sister vessels.

The heavy cruiser was laid down in the Deschimag dockyard in Bremen in August 1937 and commissioned in July 1939. As fitting out of the vessel unfolded, Germany sold the still-incomplete *Lützow* to the Soviet Union in the wake of the Molotov–Ribbentrop pact. On 15 April, the vessel was towed to Leningrad, where it was renamed *Petropavlovsk*.

Lützow was designed when complete to have a standard displacement of 17,600 tonnes (17,300 tons) and mounted eight 20.3cm (8in) SK C/33 L/60 guns. Confusingly, after the sale, the Germans reclassified as a heavy cruiser the pocket battleship *Deutschland* and renamed her *Lützow*.

Light Cruiser *Emden*

Emden was the first major warship constructed by the Reichsmarine after the end of World War I. She had a standard displacement of 5400 tonnes (5300 tons). Her main armament consisted of eight turrets, each with a single 1916-vintage 15cm (5.9in) SK C/16 L/45 gun; this unusual arrangement was dictated by compliance with the Treaty of Versailles. In 1942, these old weapons were replaced with more modern 15cm (5.9in) SK C/36 guns that could achieve an enhanced maximum range of 23,300m (76,443ft).

Emden spent most of her pre-war service history as a naval cadet-training ship. During this period, she also underwent several major modification programmes, including one that replaced her coal-fired boilers with oil-fired ones. *Emden* next participated in the April 1940 German invasion of Norway, forcing a passage through the Oslofjord with the ill-fated *Blücher* and *Lützow*. Next, in September 1941, she provided fire support for German Army operations against the Soviets in the Gulf of Riga.

Subsequently, in 1942–44, she either undertook training missions in the Baltic Sea or mine-laying operations off the Norwegian coast. Finally, between 1 and 7 February 1945, the barely-operational ship sailed at reduced speed to Kiel's dockyard for major repairs. After Allied air attacks had badly damaged her, leaving her with a significant list, she was towed to the nearby Heikendorfer Bay and ran aground. On 3 May, her crew scuttled the vessel to avoid her being captured by the advancing British forces.

Emden

This view of the light cruiser *Emden* in her training role clearly shows her two tall narrow funnels amidships and the long relatively shallow bridge structure.

Emden

Displacement: 7100 tonnes (6990 tons) full load

Dimensions: 155.1 x 14.2 x 5.3m (508ft 10in x 46ft 7in x 17ft 5in)

Propulsion: steam turbines, 2 shafts, 10 boilers, 34,700kW (46,500hp)

Armament: 8 × 15cm (5.9in) SK L/45 guns; 3 × 8.8cm (3.4in) SK L/45 guns; 4 × 50cm (20in) torpedo tubes

Armour: belt: 50mm (2in); armour deck: 40mm (1.57in); conning tower: 100mm (3.94in)

Speed: 29.5 knots (55km/h; 34mph)

Crew: 30 officers, 445–653 enlisted men

***Königsberg* in docks**
This view of the stern of the light cruiser *Königsberg*, berthed in a northern German port, illustrates the sheer length of the vessel's aft hull and superstructure, on which were mounted two of the ship's three treble 15cm (5.9in) SK C/25 gun turrets.

Königsberg
Displacement: 7800 tonnes (7700 tons) full load
Dimensions: 174 x 15.3 x 6.28m (571ft x 50ft x 20ft 6in)
Propulsion: 2 × MAN 10-cylinder diesel engines; 4 × geared steam turbines; 3 × shafts
Armament: 9 × 15cm (5.9in) SK C/25 guns; 2 × 8.8cm (3.4in) SK L/45 anti-aircraft guns; 12 × 50cm (1.96in) torpedoes; 120 mines
Armour: belt: 50mm (1.96in), armour deck 40mm (1.57in), conning tower 100mm (3.9in)
Speed: 32 knots (59km/h; 37mph)
Crew: 21 officers, 493 enlisted men

Light Cruiser *Königsberg*

Königsberg was the first ship in this three-vessel class of light cruisers. She was laid down in the Germaniawerft at Kiel in April 1926 and was commissioned on 17 April 1929. Her full-load displacement was 7800 tonnes (7700 tons). Her armoured protection varied from 40mm (1.57in) on her deck through to 100mm (3.9in) on her conning tower. Her main armament consisted of nine 15cm (5.9in) SK C/25 guns, with one treble turret forward and two offset super-firing treble turrets aft. *Königsberg* carried 120 rounds for each of these nine guns. Her secondary armament included two 8.8cm (3.4in) SK L/45 flak guns.

In her early Reichsmarine service, *Königsberg* served as a cadet-training vessel. Then she had a seaplane catapult and crane installed. Next, in April 1940, she participated in the German invasion of Norway. She formed part of Group 3 and transported 600 infantrymen of the 69th Division to Bergen. *Königsberg* stormed into Bergen harbour, but was hit forward three times by Norwegian coastal artillery; these caused significant flooding and left the vessel barely manoeuvrable.

On 10 April, 16 Blackburn Skua dive-bombers attacked the anchored light cruiser and hit her five times with 45kg (100lb) bombs. These caused significant damage; the ship's list increased until she eventually capsized and sank. The vessel's operational career had lasted just three days.

Firepower on *Karlsruhe*, 1940	
Main armament	9 x 15cm (5.9in) SK C/25 guns
Secondary armament	2 x single 8.8cm (3.4in) SK L/45 flak guns
Torpedo tubes	4 x treble tubes amidships
Torpedoes carried	24 x 50cm (20in) torpedoes
Naval mines	120

Light Cruiser *Karlsruhe*

The second vessel in the class, *Karlsruhe*, featured the same loaded displacement and armament as her sister ship *Königsberg*. She was powered by four geared steam turbines that enabled her to reach a maximum operating speed of 32 knots (59km/h; 37mph). Her protection included a 50mm (1.96in) main armoured belt and 40mm (1.57in) of deck armour. During the 1930s, *Karlsruhe* undertook several overseas naval cadet-training cruises and underwent three major refits. These modifications enlarged her stern, fitted both her funnels with raked caps and replaced the 8.8cm (3.4in) guns with 10.5cm (4.13in) ones.

Her first major mission commenced on 8–9 April 1940, when she participated in the German invasion of Norway. She sailed from Bremerhaven with the mission of capturing the port of Kristiansand. After securing the port, *Karlsruhe* steamed back out of the fjord, where she was hit by two torpedoes fired by the British submarine HMS *Truant*. The strikes caused significant flooding and put her engines out of action. The Germans abandoned the crippled vessel and the nearby German torpedo boat *Greif* fired two more torpedoes to ensure that she sank. At just two days, the combat history of *Karlsruhe* was even shorter than that of *Königsberg*.

Light Cruiser *Köln*

The final vessel in this class, *Köln* was laid down at Kiel during August 1926. Launched in May 1928, she was finally commissioned on 15 January 1930. She spent that year working up and had her single 8.8cm (3.4in) anti-aircraft guns replaced with double mounts. In the 1930s, she underwent the same modifications as those incorporated in her sister vessel *Karlsruhe*.

Köln participated in the invasion of Norway in early April 1940. She stormed Bergen alongside her sister ship *Königsberg*, incurring modest damage. Subsequently in 1940, she went into dry dock and had an experimental landing pad for the Flettner FL 272 helicopter erected on top of her 'B' turret. After service in 1941–42 in Baltic and Norwegian waters, as well as two periods in dry dock, she was decommissioned on 17 February 1943 at Kiel. In July 1944, she came out fully operational from dry dock and served in Norwegian waters. Damaged in Oslofjord by British bombers, she returned to Wilhelmshaven for repairs. On 30 March 1945, US bombers hit her twice, causing her to sink upright in shallow waters. Amazingly, with her guns still above water, she was able to provide fire support for the German ground forces defending the approaches to Wilhelmshaven during the war's final few days.

Light Cruiser *Leipzig*

The lead ship in a two-vessel class, *Leipzig* was laid down in April 1928 and commissioned in October 1931. Featuring three treble 15cm (5.9in) gun turrets, the ship displaced 8100 tonnes (8000 tons) unloaded and could obtain a top cruising speed of 32 knots (59km/h; 37mph).

In late 1939, *Leipzig* undertook mining missions in the North Sea before being torpedoed by a British submarine and suffering extensive damage; she spent most of 1940 in dry-dock repair in Kiel. In summer 1941, she provided fire support during the Axis invasion of the Soviet Union. Thereafter, *Leipzig* undertook cadet-training duties in the Baltic in 1942–43, followed by another major refit.

In autumn 1944, she joined the Baltic Fleet, providing fire support to the German ground forces that were attempting to stem the Soviet advance westward. On 14 October 1944, while sailing in thick fog, she collided with *Prinz Eugen*, and was almost cut in two. After being towed to Gotenhafen, it

Flettner FL 272 helicopter	
Number built	24
Service	1942–45
Engine	120kW (150hp) Siemens-Halkse Sh 14
Missions	Reconnaissance; submarine spotting
Maximum speed	150km/h (93mph)

Flettner FL 272 helicopter
The Flettner was extensively test-flown in the ship-borne reconnaissance role throughout 1941, operating from a small helicopter pad from the cruiser *Köln*. The Kreismarine were impressed with its perfromance, ordering 15 prototypes and 30 production machines. By 1943, more than 20 were serving in the Baltic, Aegean and Mediterranean.

Leipzig

The profile of the two Leipzig-class vessels stood out in comparison with all other German light cruisers by having just one large raked funnel stack rather than two narrower stacks.

Leipzig

Displacement: 8100 tonnes (8000 tons)
Dimensions: 177 x 16.3 x 5.69m (580ft 9in x 53ft 6in x 18ft 8in)
Propulsion: steam turbines and diesels; 3 shafts (diesels on centre shaft); 45MW (60,000shp) turbines + 9.3MW (12,400hp) diesels
Armament: 9 × 15cm (5.9in) SK C/25 guns; 2 × 8.8cm (3.4in) SK L/45 anti-aircraft guns; 12 × 50cm (19.6in) torpedoes; 120 mines
Armour: belt: 50mm (1.96in); armour deck: 30mm (1.2in); conning tower: 100mm (3.9in)
Speed: 32 knots (59km/h; 37mph)
Aircraft carried: 3 x Arado Ar 196 seaplanes
Crew: 26 officers, 508 enlisted men

was decided that her damage was too extensive to repair fully. She was merely patched up so that she could remain afloat when static. She provided fire support as the Red Army closed in around the port.

Finally, on 24 March, and packed full of evacuees, *Leipzig* limped to the port of Hela and thence to Denmark, which she reached on 29 April. She was in such a bad condition she played no further part in operations.

DESTROYERS

The effective operation of German capital ships and cruisers was facilitated by escorting destroyers. These vessels provided protection screens for larger ships, as well as carrying out screening and reconnaissance roles. Acting independently, destroyers could also carry out mine-laying missions or provide fire-support for ground forces in littoral areas. Germany produced 40 operational destroyers before and during the 1939–45 war. This section will look at the six main German destroyers classes: the four Type 34 vessels; the 12 Type 34As; the six Type 36 ships; the eight Type 36As; the seven Type 36A(Mob) vessels; and the three Type 36Bs.

Type 34 destroyers (Z1–Z4)

These four named boats (Z1–Z4) were laid down in 1934–35, launched in 1935, and commissioned in 1937. The design incorporated new 51,453kW (69,000hp) Wagner steam-powered turbines, which enabled the vessel to reach a maximum speed of 36 knots (67km/h; 41mph). This propulsion system, however, subsequently proved unreliable.

Displacing 3206 tonnes (3155 tons) fully loaded, the destroyers' main armament comprised five single 12.7cm (5in) SK C/34 guns; two of these were located forward, and three aft. The ships' secondary weaponry included four 3.7cm (1.46in) and six 2cm (0.79in) Flak guns, eight torpedo tubes and 60 sea mines.

These four ships spend most of 1938–39 undertaking training exercises. Then, during 19 February 1940, the First Destroyer Flotilla undertook a sortie into the North Sea. The flotilla comprised three Type 34s (Z1, Z3 and Z4) and three other destroyers (Z6, Z13 and Z16). While passing through a cleared passage in German minefields, Z1 was attacked by a Heinkel He 111 bomber, which mistook her for a British warship. The bomber scored three hits, which caused Z1 to break in two. When her sister ship Z3 went to rescue the many survivors in the water, she exploded and quickly sank, probably after hitting a British mine.

Z2 was the only Type 34 not to participate in the North Sea sortie. She did, however, form part of the Escort

Leberecht Maas
Displacement: 3206 tonnes (3155 tons) full load
Dimensions: length: 119m (390ft); beam: 11.3m (37ft)
Propulsion: 2 x steam turbines, 52,199kW (70,000hp)
Armament: 5 x 127mm (5in); 2 x twin 37mm (1.46in) AA; 6 x 20mm (0.79in) AA; 2 x quadruple torpedo mounts; 60 mines
Speed: 38 knots (70km/h; 44mph)
Crew: 315

Leberecht Maas
In this starboard profile of the Type 1934 destroyer, Z1 *Leberecht Maas*, the ship's torpedo tubes can be seen amidships, as well as one of its life-boats, located adjacent to the forward funnel.

Group One that covered the April 1940 German invasion of Norway, specifically the assault on Narvik. During an intense engagement with five British destroyers, Z2 was hit seven times and suffered extensive damage; her forward gun and fire control system were destroyed, one magazine flooded and several large fires raged. On 13 April, she suffered more damage at the hands of British destroyers; the crippled vessel was run aground so that her crew could disembark. The stricken vessel eventually broke up.

Z4, in contrast, had a long wartime career. She participated in the February 1942 Channel Dash and the 31 December 1942 Battle of the Barents Sea. She spent much of 1943 operating in Norwegian waters. She was damaged by Allied aircraft in April 1945 and remained under repair until the war's end.

Type 34A destroyers (Z5–Z16)

The Type 34A was a slightly modified variant of the preceding design. Twelve named vessels (Z5–Z16) were laid down during 1935 and commissioned in 1937–39. Their specifications and combat performance were identical to the Type 34s. Eight of these destroyers (Z5–6, Z8–Z9, Z11–Z13, Z16) participated in the April 1940 German invasion of Norway; only four survived the operation.

On 10 April, during the First Battle of Narvik, five British destroyers engaged eight German escorts. Subsequently, during the Second Battle of Narvik on 13 April, Z9 attempted to torpedo the battleship HMS *Warspite*, while Z13 was reduced to a blazing wreck by British destroyer fire; her crew scuttled the vessel. Thereafter, the surviving German warships were hemmed in by the Royal Navy. Having run out of ammunition, Z9 and Z11 were scuttled in the Rombaksfjord. Subsequently, Z12, crippled with engine failure, was hit 21 times by British fire and eventually sank.

The eight surviving Type 34As conducted escort and mine-laying duties throughout 1941 without further loss. However, in late January 1942, Z8 was speeding through the English Channel when she hit two mines and sank. Subsequently, on 30 April 1942, Z7, along with Z24 and Z25, sailed from northern Norway to intercept Convoy QP11.

On 2 May, the German destroyers encountered the British light cruiser HMS *Edinburgh*, which had been crippled by a U-boat torpedo. The British cruiser scored two hits that disabled Z7, after which she was scuttled.

Subsequently, Z16 participated in the 31 December 1942 Battle of the Barents Sea, escorting *Admiral Hipper*. Surprised by the British cruiser HMS *Sheffield*, Z16 was hit five times in two minutes and broke in half. The five remaining Type 34s (Z5, Z6, Z10, Z14 and Z15) served out the rest of the war. In 1944–45, Z5 evacuated German wounded military personnel and civilians as the Red Army advanced through Prussia and Pomerania.

Finally, in the war's last three weeks, three Type 34As (Z6, Z10 and Z14) left Norway for the Baltic to rescue as many Germans as possible from the East to avoid them falling into Soviet captivity.

Type 36 destroyers (Z17–Z22)

The Type 36 design, of which six named vessels were constructed, was an enlarged version of the previous two classes that incorporated modifications. These six vessels were laid down in Bremen in 1936–38 and were commissioned between August 1938 and September 1939. These vessels' standard displacement was 2450 tonnes (2411 tonnes). They featured redesigned bows that made them better 'sea-boats' than the previous Type 34 and 34A vessels.

During the autumn of 1939–40, these destroyers undertook missions to support the invasion of Poland as well as North Sea mining operations. Then, during April 1940, five ships in this class (Z17–19, Z21–22) participated in the German invasion of Norway, serving as escorts for the attack on Narvik. During the 10 April 1940 First Battle of Narvik, fought out in the Ofotfjord, five British H-class destroyers caught Z21 and Z22 by surprise in the Rombaksfjord. After a fierce encounter, British fire sank both destroyers. Three days later, a British naval task force once again entered the Ofotfjord.

In the ensuing Second Battle of Narvik, the German destroyers fought until they ran out of ammunition. The undamaged Z19 retreated to the Herjangsfjord, while Z18 withdrew to the Rombaksfjord; both ships were then scuttled. Z17 was sunk by British fire while still docked in Narvik's harbour. By 13 April, therefore, five of the six vessels of this class had been lost.

The sole Type 36 survivor, Z20, saw operational service through until the end of the war, serving in Norwegian waters throughout 1941. During early 1945, Z20 took part in the evacuation of

Z20 *Karl Galster*
On 8 May 1945, Z20 joined the last desperate efforts to rescue troops from the eastern end of the Hel peninsula north of Danzig in Prussia. Crammed with 2000 passengers she sailed west just hours before Germany surrendered unconditionally; some of those unlucky not to get on board desperately attempted to paddle their way westward to safety on rafts.

Z20 Karl Galster (Type 36 destroyer)
Displacement: 2450 tonnes (2411 tons)
Dimensions: length: 123.4m (404ft); beam 11.8m (38ft 9in)
Propulsion: 51,000kW (69,000hp); 6 × water-tube boilers
Armament: 5 × single 12.7cm (5in) guns; 2 × twin 3.7cm (1.46in) anti-aircraft guns; 7 × single 20mm (0.79in) C/30 AA guns; 2 × quadruple 53.3cm (21in) torpedo tubes; 60 mines
Speed: 36 knots (67km/h; 41mph)
Crew: 323

German civilians and wounded military personnel from the parts of Prussia and Pomerania threatened by the advance of the Red Army.

On 5 May, as northwestern Germany capitulated, the new president, Grand-Admiral Dönitz, ordered all available naval assets to race eastward through the Baltic Sea to rescue as many Germans as possible from the enclaves of the Hela Peninsula and Courland before Germany unconditionally surrendered. Z20 obeyed Dönitz's call and raced to Hela. On 8 May, she took on board 2000 soldiers and raced westward to surrender to the British in Flensburg on 10 May.

Type 36A 'Narvik' destroyers (Z23–Z30)

As designed, this group of eight unnamed destroyers were intended to feature three single potent 15cm (5.9in) guns turrets aft with a double turret in the bow; the 15cm (5.9in) calibre was usually associated with German light cruisers rather than destroyers. All eight vessels, however, were initially built with just a single forward turret due to shortages in the twin turret; subsequently, four vessels in the class (Z23–25 and Z29) had the single fore turret replaced with a double turret; this brought their total main armament configuration to five (rather than four) powerful 15cm (5.9in) gun barrels. These vessels thus packed greater firepower than any of their predecessor destroyers.

The ships had a standard displacement of between 2643 tonnes (2601 tons) and 2700 tonnes (2657 tons). Their geared steam turbines generated 52,199kW (70,000hp), which enabled them to obtain an impressive maximum speed of 37.5 knots (69km/h; 43mph). Despite the lessons garnered from previous designs, these vessels were not regarded as good 'sea boats' when in heavy seas; the weight of the double forward turret exacerbated this problem.

Z26

Z26 was one of four Type 36A destroyers that retained their original forward-located single 15cm (5.9in) guns; her two sets of quadruple torpedo tubes can be seen amidships, either side of the second funnel.

Z26 (Type 36A destroyer)
Displacement: 2700 tonnes (2657 tons)
Dimensions: length: 127m (417ft); beam: 12m (39ft)
Propulsion: 2 x steam turbines, 52,199kW (70,000hp)
Armament: 4 x 150mm (5.9in) guns; 4 x 3.7cm (1.46in) SK C/30; 8 x 20mm (0.79in) C/30; 8 x 533mm (21in) torpedo tubes
Speed: 37.5 knots (69km/h; 43mph)
Crew: 320

Z25, Type 1936A destroyer
Z25 was just one of three vessels of the Type 1936A class that survived the war; subsequently she served in the French Navy as *Hoche* until decommissioned in 1956; she was scrapped in 1961.

These eight ships were laid down at AG Weser's shipyard in Bremen in 1938–40 and were commissioned between September 1940 and November 1941. During 1943–45, all the surviving vessels incorporated augmented flak defences, including additional 3.7cm (1.46in) and 2cm (0.79in) weapons.

Just three ships of this class (Z25, Z29 and Z30) survived the war. On 29 March 1942, during an action in the Barents Sea off northern Norway, destroyer Z26 was sunk by British cruiser and destroyer gunfire. Next, during 28 December 1943, a force of five German destroyers and six torpedo boats attacked two British light cruisers; in the ensuing three-hour Battle of the Bay of Biscay, British fire sank Z27.

The next 'Narvik' Type lost was Z23. She was badly damaged by Allied bombers on 12 August 1944 and was stricken thereafter. Some 12 days later, Allied bombers also sank Z24 in the Gironde estuary. The final vessel lost was Z28, which was sunk by Allied bombers in Sassnitz harbour on 6 March 1945. Z30 was badly damaged in Oslofjord in October 1944 and remained non-operational until the war's end.

Type 36A (Mob) destroyers (Z31–34, Z37–Z39)

These seven vessels were similar to the preceding class. All bar one of them (Z31) were built with a twin 15cm (5.9in) forward gun turret, giving them five such barrels; this extra weight forward, however, did not help their seaworthiness. During 1944, Z31 had her single forward turret removed and replaced with a double turret; when under repair in early 1945, Z31 had this turret replaced with a single 10.5cm (4.13in) gun.

These seven vessels were laid down in 1940–1941 and commissioned between 11 April 1942 and 21 August 1943. The AG Weser shipyard in Bremen built four of these destroyers (Z31–34); the Germaniawerft yard in Kiel built the remaining three. These ships displaced 2645 tonnes (2603 tons) and with a full fuel load could obtain an operational range of 5127 nautical miles (5900km; 5901 miles).

In 1943, Z32 and Z37 participated in operations to facilitate the efforts of 'blockade runners' to break out from, or break back into, French Atlantic coast ports. Z36 was also involved in the Battle of the Bay of Biscay on 28 December 1943, but survived with only modest damage. On 30 December, Z37 collided with Z32 in the Bay of Biscay; she suffered such extensive damage that, after being towed back to Bordeaux, she was considered beyond repair. Her guns were removed and she was scuttled on 24 August.

On 9 June 1944, three days after the D-Day landings, three German destroyers, including Z32, were moving from Brest to Cherbourg when they were intercepted by eight Allied destroyers. During the ensuing Battle of the Ushant, Z32 was crippled by gunfire and was driven ashore on the Ile de Batz, where she was scuttled.

The other five vessels (Z31, Z33, Z34, Z38, Z39) saw service until the war's end. Z31 spent the whole of 1943 operating in Norwegian waters, escorting larger ships in a series of major sorties. After an eight-month refit in Wesermünde, Z31 then returned to Norwegian waters. Finally, along with Z34 and Z38, she transferred to the Baltic to provide fire support to the German ground forces.

Z33, Z34 and Z38 spent most of 1943–44 operating in Norwegian waters before the latter returned to the Baltic in early 1945 to both evacuate German personnel from the east and provide fire support. On 7–9 May, Z38 and

Z38

After her last-gasp efforts during the war's final hours to rescue German troops on the Hela peninsula from imminent Soviet captivity, Z38 was taken into Royal Naval service as HMS *Nonsuch* (D107).

Type 36A (Mob) destroyer
Displacement: 2645 tonnes (2603 tons)
Dimensions: length: 127m (417ft); beam: 12m (39ft); draught: 4.62 m (15 ft 2 in)
Propulsion: 2 x steam turbines, 51,000kW (69,000hp)
Armament: 5 × 15cm (5.9in) guns; 4 (later 14) × 3.7cm (1.45in) guns; 12 (later 18) × 2cm (0.79in) guns; 8 × 53cm (21in) torpedo tubes; 60 mines; 4 × depth charge launchers
Speed: 37.5 knots (69.5km/h; 43.2mph)
Crew: 330

Destroyer names			
Vessel	**Type**	**Vessel**	**Type**
Z1 *Leberecht Maas*	34	Z12 *Erich Giese*	34A
Z2 *Georg Thiele*	34	Z13 *Erich Koellner*	34A
Z3 *Max Schultz*	34	Z14 *Friedrich Ihn*	34A
Z4 *Richard Beitzen*	34	Z15 *Erich Steinbrinck*	34A
Z5 *Paul Jacobi*	34A	Z16 *Friedrich Eckoldt*	34A
Z6 *Theodor Riedel*	34A	Z17 *Diether von Roeder*	36
Z7 *Hermann Schoemann*	34A	Z18 *Hans Lüdemann*	36
Z8 *Bruno Heinemann*	34A	Z19 *Herman Künne*	36
Z9 *Wolfgang Zenker*	34A	Z20 *Karl Galster*	36
Z10 *Hans Lody*	34A	Z21 *Wilhelm Heidkamp*	36
Z11 *Bernd von Arnim*	34A	Z22 *Anton Schmitt*	36

Z39 participated in the final, frantic, personnel evacuations from the Hela Peninsula.

Type 36B destroyers (Z35–36, Z43–45)

The negative impact on seaworthiness caused by incorporating a forward twin 15cm (5.9in) turret in the Type 36A(Mob) class led the Navy to mount five single 12.5cm (5in) SK C/36 L/45 gun turrets in the final class built during the war, the Type 36B. These vessels displaced 3600 tonnes (3540 tons) fully loaded and were powered by two 26,099kW (35,000hp) Wagner geared turbines, which enabled them to reach a top speed of 36.5 knots (68km/h; 42mph). Eight ships were initially ordered, but three (Z40–Z42) were cancelled. The remaining five ships (Z35–36, Z43–45) were laid down in Bremen in 1941–42. Three vessels, Z35, Z36 and Z43, were commissioned between September 1943 and March 1944. The fourth ship in the class, Z44, was close to being commissioned when, on 29 July 1944, she was badly damaged in an Allied bombing raid; she remained non-operational until the war's end. The final vessel, Z45, was only partly complete when the decision was taken during late 1944 to suspend work on her.

Z35, Z36 and Z43 served with the Sixth Destroyer Flotilla in the Baltic Sea during most of 1944. On 12 December 1944, these vessels participated in the laying of a minefield off the shores of Estonia; Z35 and Z36 each

Type 23 torpedo boat
This image of the Type 23 torpedo boat *Seeadler* nicely illustrates the design's larger front funnel and modestly-sized bridge; it would appear that the boat is speeding at right angles through a column of larger German surface vessels.

inadvertently hit a German mine and sank. In early 1945, Z43 provided naval gunfire support in the Kolberg area until it too hit a mine on 10 April and was rendered unserviceable; her crew scuttled the vessel on 3 May.

SMALL COMBAT VESSELS

Although the destroyer was the principal Kriegsmarine escort vessel, other small combat vessels also played a key role in German maritime operations. This section examines the four most important types of such vessels: T-class torpedo boats, S-class E-boats, R-boats and M-class minesweepers.

Torpedo boats were fast vessels with a standard displacement in the 600–1900 tonne (590–1869 ton)

Seeadler (Type 23 torpedo boat)
Displacement: 1290 tonnes (1270 tons) full load
Dimensions: length: 87.7m (288ft); beam: 8.43m (27.7ft)
Propulsion: 2 x steam turbines, 17,151kW (23,000hp)
Armament: 3 x 105mm (4.13in); 7 x 20mm (0.79in) AA; 6 x 533mm (21in) torpedo tubes; 30 mines
Speed: 33.6 knots (62km/h; 39mph)
Crew: 120–129

range that were equipped with guns and six torpedo tubes. The smaller E-boat motor torpedo boat (MTB) displaced around 100 tonnes (98 tons) and sported two torpedo tubes. Displacing 60–160 tonnes (59–157 tons), R-boats were multi-role motor launches that often acted as auxiliary minesweepers. With a displacement of 550–725 tonnes (541–713 tons), the larger M-class vessels were dedicated minesweepers.

Torpedo boats

During the war, the Germans operated 69 torpedo boats, including 14 vintage ships from World War I; six named 'Raubvogel' vessels; six 'Raubtier' boats; 12 unnamed Type 35s (T1-12); 12 Type 37s (T13-T21); 15 Type 39 fleet boats (T22-36); and four captured boats. Entering service in 1926–27, the six 'Raubvogel' vessels (including Kondor) were enlarged versions of extant World War I vessels.

Their standard displacement was 939 tonnes (924 tons), their top speed was 33 knots (61km/h; 38mph) and their armament included three 10.5cm (4.13in) guns and six 53cm (21in) torpedo tubes. All six boats were lost during the war; one was sunk by gunfire during the 1940 invasion of Norway; one was sunk by a British MTB in mid-1942; and four were lost due to Allied bombing in summer 1944.

Commissioned in 1929, the six 'Raubtier' vessels had a standard displacement of 950 tonnes (934 tons) and mounted six torpedo tubes plus either three 10.5cm (4.13in) or three 12.7cm (5in) guns. During 1939–44, all six vessels were lost. Launched in 1937–39, the 12 Type 35 boats (T1–12) were smaller, faster, versions of the earlier classes, which mounted six torpedo tubes but only one 10.5cm (4.13in) gun. Five Type 35s were sunk by Allied bombing and two by Allied mines.

The 12 Type 37s constructed in 1940–42 (T13–T21) were similar to the Type 35s, but with 3.7cm (1.46in) Flak guns instead of 2cm (0.79in) ones. Subsequently, in response to the growing Allied air threat, these boats incorporated up to 10 additional 2cm (0.79in) Flak guns. During the war, four of these vessels (T13, T15, T16 and T18) succumbed to Allied bombing. The 15 Type 39 'Elbing' fleet boats (T22–36) had a standard displacement of 1315 tonnes (1294 tons) and sported six torpedo tubes; they also mounted four 10.5cm (4.13in) guns and nine smaller-calibre Flak weapons.

Six Type 39s participated in the December 1934 Battle of the Bay of Biscay; British fire sank both T25 and T26. Four months earlier, T20, T30 and T32 sank after hitting German mines in the Bay of Finland. This class's most notable success was the sinking of a British light cruiser in 1943.

E-Boat fast attack craft

The German S-class Schnellboot fast attack craft (or E-Boat motor torpedo boat; 'MTB' in Allied parlance) was a small, fast vessel that mounted two torpedo tubes, mine-laying devices and small-calibre Flak guns, which operated in coastal waters. A typical E-Boat displaced around 100 tonnes

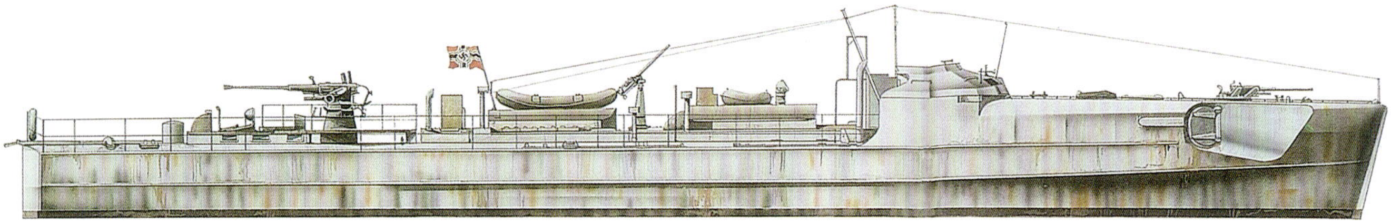

S-Boot

The S-139 sub-class of the German S-Boat fleet featured a shallower profile than its predecessor designs thanks to the installation of a significantly lower armoured bridge.

S-Boot
Displacement: 80.17 tonnes (78.9 tons) full load
Dimensions: 32.76 x 5.06 x 1.47m (107ft 6in x 16ft 6in x 4ft 10in)
Propulsion: 3 × Daimler Benz MT 502 marine diesel engines, 2950kW (3960hp)
Armament: 2 x 533mm (21in) torpedo tubes (4 torpedoes); 3 × twin 20mm (0.79in) C/30 cannon; 1 × 37mm (1.46in) Flak 42 cannon
Speed: 43.8 knots (81km/h; 50mph)
Crew: 24–30

(98 tons), could reach a speed of 40–44 knots (74–81km/h; 46–50mph) and had a maximum operational radius of 1481km (920 miles).

There were three main types of E-Boat: the early low-forecastle boats (including the S1, S7, S14 batches); the mid-war high-forecastle boats (the S26 and S30 classes); and the late-war low armoured bridge boats (the S139 and S170 sub-classes). The seven S7 boats (S7–S13), the first standardized batch, had a standard displacement of 86 tonnes (84 tons) and mounted a single 2cm (0.79in) Flak 30 gun. The four S14 boats (S14–17), in contrast, displaced 100 tonnes (98 tons) and were powered by two MAN 11-cylinder diesel engines that proved unreliable.

The mid-war class of boats (classes S26, S30 and S100) had their forecastle (the boat's front hull) raised so that it enclosed the torpedo tubes. The 91-strong S26 class displaced 100 tonnes (98 tons); powered by three DB 20-cylinder diesel engines, these vessels obtained a maximum speed of 41 knots (76km/h; 47mph).

The 16 units of the S30 class (S30–37, S54–61) were powered by three smaller DB 16-cylinder diesel engines and could reach a top cruising speed of 36 knots (67km/h; 41mph). They carried six torpedoes (instead of their predecessor's four) and also mounted one 2cm (0.79in) Flak gun towards the stern, together with two machine guns.

The S100 class were powered by three DB MB-501 diesel engines and could obtain a maximum speed of 44 knots (81km/h; 50mph); these boats mounted four 3.7cm (1.46in) or 2cm (0.79in) Flak guns. The final type of E-Boat constructed featured a redesigned shallow armoured bridge structure that reduced the vessel's overall side profile. The S139 and S170 sub-classes were 36m (118ft) in length and were powered by three supercharged DB MB-511 engines.

E-Boats operated in many coastal waters, including the Norwegian coast, the Baltic, the Channel, the coastal waters of occupied France, the Mediterranean and the Black Sea. Throughout the war, E-Boats sank 101 merchant ships, 12 Allied destroyers and 28 other naval vessels; the mines these E-Boats laid sank 38 merchant ships and seven warships. The Germans lost 83 E-Boats in the Channel alone.

SUBMARINES

Alongside the surface vessels already discussed, the other weapon that spearheaded Kriegsmarine wartime operations was the submarine. These vessels were termed Unterseeboote ('underwater boats') or 'U-boats'. Rather than names, each of the 1156 submarines commissioned was given a 'U' prefix and a unique number from 1 through to 4712.

The savage sustained struggle under the surface waters during the Battle of the Atlantic, interdicting supplies desperately needed by the Allies, formed a critical dimension of the entire war. With U-boat combat deployment peaking with over 100 boats at sea most days between August 1942 and July 1943, German submarines sank over 14 million tonnes (13.7 million tons) of shipping during the war.

This section examines the six main U-boat models: Type I and Type II; the various Type VIIs; the VIIc/41 boats; the Types XXI and XXIII 'Electro-boats'; and finally a miscellany of specialist types.

Type I and Type II U-boats

The two Type IA German submarines were the Navy's first attempt to develop an ocean-going combat boat. Laid down in summer 1935, these two boats were commissioned in spring 1936. They displaced 981 tonnes (966 tons) submerged and featured two MAN diesel engines and two double-acting electric motors. These boats could

Kiel harbour

Three pre-war Type VIIA submarines, including U-27, U-33, and U-34, lie berthed near the 'Blücher Bridge' in Kiel dockyard sometime during the first half of 1939. In the background (left) lie two Type IIA submarines.

obtain an underwater operational range of just 144km (90 miles) when moving at 4 knots (7km/h; 5mph), but at 14,600km (9100 miles), their surface range was impressive. These boats proved to be difficult to navigate and slow in the dive. Between them, these boats sank 18 Allied merchant ships.

In 1934–40, the Germans also constructed 50 small Type II coastal patrol submarines in four sub-variants (IIA, IIB, IIC and IID). The IIA design displaced 310 tonnes (305 tons) when submerged and had a limited underwater maximum range of just 55km (34 miles) when moving at 4 knots (7km/h; 5mph); the top submerged speed was 6.9 knots (13km/h; 8mph).

The boats carried five torpedoes, delivered from three bow tubes. The 16 slightly heavier Type IID boats displaced 314 tonnes (309 tons) submerged and could achieve a maximum underwater movement range of 104km (64 miles) at 4 knots (7km/h; 5mph). Originally intended primarily for training, they were also employed from 1940 onwards for combat patrols in British coastal waters.

Type VIIA, VIIB and VIIC U-boats

The Type VII medium-range diesel-electric combat submarine was the most common wartime U-boat, with 590 Mark A, B and C boats of being produced. In 1935–37, the Germans constructed 10 Type VIIA boats. They were powered by two 179kW (240hp) MAN diesel engines and, when submerged, by two 279kW

U-2

Constructed at Kiel, U-2 was commissioned during 1935. After undertaking two combat patrols during 1940, the boat was employed as a 'school boat' in the 21st Training Flotilla at Pillau in East Prussia.

U-2 (Type IIA)

Displacement: 310 tonnes (305 tons) submerged
Dimensions: length: 43.9m (144ft); beam: 4.8m (16ft)
Propulsion: surfaced: diesels 522kW (700hp); submerged: electric motors 306kW (410hp)
Armament: 3 x 533mm (21in) torpedo tubes; 1 x (later 4) 20mm (0.79in) AA gun
Range: surfaced: 2575km (1600 miles) at 8 knots (15km/h; 9mph); submerged: 55km (34 miles) at 4 knots (7km/h; 5mph)
Speed: surfaced: 13 knots (24km/h; 15mph); submerged: 6.9 knots (13km/h; 8mph)
Crew: 22–24

Type VIIAs in port

Three pre-war Type VIIA submarines, U-30, U-31 and U-32, tied up at a dockyard; note the large white identification number on the conning tower – this was removed when war came. The 88mm (3.4in) flak gun mounted on this class can clearly be seen on the deck of U-32.

(375hp) double-acting electric motors. These submarines could achieve a top underwater speed of 7.6 knots (14km/h; 9mph) and a submerged range of 150km (92 miles) when travelling at a steady 4 knots (7km/h; 5mph). They had four bow and one stern tubes for which they carried 11 torpedoes. In addition, the design mounted an 8.8cm (3.4in) gun on the upper deck. The design also had a maximum crash depth of 220m (722ft). Allied activity sank all but two of these boats.

Built in 1936–41, the 24 Type VIIB boats completed featured increased operational range. The design incorporated external fuel tanks that extended their surface operational range from 10,000km (6214 miles) to 14,000km (8699 miles). They also carried three extra torpedoes and had two rudders rather than one. These combat boats proved effective on operations. In 1939–41, for example, U-48 undertook 12 combat patrols and sank 51 merchant ships. During the same period, U-47, which sank the battleship HMS Royal Oak in Scapa Flow, mounted 10 combat patrols and sank 30 merchantmen.

By far the most numerous U-boat variant of the war was the Type VIIC, of which 568 were constructed in 1938–44. The Type VIIC was a slightly larger and heavier version of the VIIB. These boats displaced 885 tonnes (871 tons) submerged – almost 14 tonnes (14 tons) more than the VIIB. These VIIC boats were powered by two 179kW (240hp) MAN diesel engines and when submerged by two 279kW (375hp) BBC double-acting electric motors. These submarines could achieve the same maximum underwater speed of 7.6 knots (14km/h; 9mph) and a submerged range of 150km (92 miles) when travelling at a constant 4 knots (7km/h; 5mph). The design incorporated for the

Under attack

This dramatic image captures the moment that a German U-boat, sailing on the surface of the ocean, is subjected to attack by an Allied maritime anti-submarine aircraft.

first time an active sonar device.

A proportion of the Type VIIC boats joined operations during the 'First Happy Time' (June–October 1940) when German submarines sank 282 vessels (amounting to 1.49 million tonnes of shipping). These VIIC boats also faced increasingly arduous missions as Allied anti-submarine responses increased significantly until by 1944, the U-boat campaign had all but lost the 'Battle of the Atlantic'.

Some of the most successful U-boats hailed from this class: in 1940–44, for example, U-552 mounted 15 combat patrols and sank 30 merchant ships and two warships; similarly, U-96 undertook 11 operational patrols and

Type VIIB

Many Type VIIB submarines enjoyed success hunting down enemy shipping during 1940–41; however, U-76 (seen here), was sunk on 5 April 1941 by British warships after completing just one mission.

U-76 (Type VIIB)

Displacement: 773 tonnes (761 tons) submerged

Dimensions: length: 61.7m (220ft); beam: 6.2m (20ft 4in)

Propulsion: surfaced: diesels 1588kW (2130hp); submerged: electric motors 559kW (750hp)

Armament: 5 x 533mm (21in) torpedo tubes; 1 x 88mm (3.5in) deck gun, 2 x 2cm (0.79in) guns

Range: surfaced 12,040km (7481 miles) at 10 knots (18km/h; 11 mph); submerged 151km (94 miles) at 4 knots (7.4km/h; 4.6mph)

U-333 (Type VIIC)
Displacement: 871 tonnes (857 tons) submerged
Dimensions: length: 66.5m (218ft); beam: 6.2m (20.3ft)
Propulsion: surfaced: diesels 2089kW (2800hp); submerged: electric motors 559kW (750hp)
Armament: 5 x 533mm (21in) torpedo tubes; 1 x 88mm (3.4in) deck gun; 1 x AA gun
Range: surfaced: 15,700km (9800 miles) at 10 knots (18km/h; 11mph); submerged: 150km (92 miles) 4 knots (7km/h; 5mph)
Speed: surfaced: 17.7 knots (33km/h; 20mph); submerged: 7.6 knots (14km/h; 9mph)
Crew: 44–52

Type VIIC
Constructed at Emden, U-333 was an example of the ubiquitous Type VIIC boat; she undertook 12 patrols during 1941–44 before being sunk by depth charges delivered by British warships on 31 July 1944.

sank 27 merchantmen. In 1944–45, some 130 of the surviving Type VIIC boats were retro-fitted with the Schnorchel, a breathing device that allowed submarines to run their diesel engines to recharge their electric motor batteries while remaining submerged just below the water's surface.

Type VIIC/41 U-boats

The final variant of the ubiquitous Type VII class ocean-going submarine was the Type VIIC/41 model, some 91 boats of which were constructed during 1941–45. No fewer than nine shipyards were involved in construction of these boats, including the Flender Werke (in Lübeck) and the Nordseewerke (in Emden).

These boats were broadly similar to their Type VIIC sister boats in technical specifications and performance. The VIIC/41 boats, however, did incorporate a strengthened hull. This enabled the crash depth on these boats to be increased by 10m (33ft) to 230m (750ft). Like their cousins, these boats could obtain a maximum surface speed of 17.7 knots (33km/h; 20mph) and a maximum submerged speed of 7.6 knots (14km/h; 9mph).

The design's maximum surface operational range was an impressive 15,700km (9800 miles); while underwater, it could travel for 150km (92 miles) at a steady speed of 4 knots (7km/h; 5mph).

Around 30 of the later VIIC/41 boats were constructed incorporating the Schnorchel breathing apparatus. In addition, the last 17 boats built (in Flensburg and Bremen) had the fittings for mine-laying deleted to simplify production. Moreover, nine of the last VIIC/41 boats to be produced incorporated an experimental synthetic rubber skin of void-filled 'anechoic' tiles, which reduced the boat's signature to both active and passive sonar devices.

U-1025

Similar to the standard Type VIIC, the Type VIIC/41 was made from thicker steel and could dive more deeply.

U-1025 (Type VIIC/41)

Displacement: 874 tonnes (860 tons) submerged

Dimensions: length: 67.2m (220ft); beam: 6.2m (20.3ft)

Propulsion: surfaced: diesels 2089kW (2800hp); **submerged:** electric motors 559kW (750hp)

Armament: 5 x 533mm (21in) torpedo tubes (14 torpedoes); 1 x 2cm (0.79in) quad and 2 x twin 2cm (0.79in) AA guns

Range: surfaced: 12,040km (6500 miles) at 10 knots (18km/h; 11mph); submerged: 150km (92 miles) at 4 knots (7km/h; 5mph)

Speed: surfaced: 17.7 knots (33km/h; 20mph); submerged: 7.6 knots (14km/h; 9mph)

Crew: 44

In addition, six of last VIIC/41s to be commissioned between autumn 1944 and January 1945 featured the new enhanced Balkongerät passive sonar hydrophone. This array consisted of 48 small hydrophones positioned along both sides of the boat, designed to 'listen' for the sound of enemy vessels in the vicinity; importantly, this device worked effectively even when the boat was merely at periscope depth.

Many of these Type VIIC/41 boats carried out combat patrols in 1943–44, when ever-increasingly effective Allied anti-submarine counter-measures inflicted an extremely heavy toll of those operational U-boats deployed at sea. The mortal danger inherent in the Battle of the Atlantic is testified by the fact that no fewer than 39 Type VIIC/41 submarines were sunk by Allied action during the war.

On watch

Four crew-members from an unidentified U-boat, wearing protective rain coats and Sou'wester rain-hats, scan the horizon with binoculars, searching for enemy vessels.

Type IX U-boats

In 1936–44, the Germans constructed some 173 Type IX large ocean-going submarines. This included eight Type IXs, 14 Type IXBs, 54 Type IXCs, 87 Type IXC/40s and 30 Type IXDs. The standard pre-war Type IX boat displaced 1170 tonnes (1152 tons) submerged and could obtain a maximum operational range of 16,900km (10,501 miles) at a constant surface cruising speed of 10 knots (18km/h; 11mph). These boats' normal diving depth was 100m (328ft) and their designed crush depth was 200m (656ft). These boats carried four bow and six stern torpedo tubes, for which they carried 22 torpedoes, as well as 10 additional ones in external containers.

The Type IXC variant, of which 87 boats were built in 1940–44, was heavier than the standard Type IX – it displaced 1251 tonnes (1232 tons) when submerged – and had a greater surface range of 22,200km (13,794 miles) when cruising at 10 knots (18km/h; 11mph). The Type IX boats spearheaded the U-boat anti-convoy operations mounted off the African and American coasts. The 14 Type IXB boats built during 1937–40 also possessed an impressive surface cruising range of 19,300km (11,992 miles). These boats delivered some of the most successful service performances of all the U-boats; each boat on average sank more than 100,000 tonnes (98,421 tons) of enemy shipping, although in the process 11 were sunk and one captured.

Type XXI 'Electro-boat' U-boats

In 1944–45, the Kriegsmarine placed excessive hope that unleashing a sustained campaign by the new generation of potent 'Electro-boat' Schnorchel-equipped streamlined submarines would wrest the strategic initiative in the war at sea back into German hands. During 1942–43, the Kriegsmarine had developed two Electro-boat designs: the large Type XXI, optimized for long-range Atlantic operations; and the small Type XXIII, intended for short-range coastal water patrols. These revolutionary new submarines featured hydrodynamically streamlined hulls and could carry

U-107

The Type IXB submarine U-107 was the fifth most successful U-boat of the war in terms of 'kills' – during 13 patrols undertaken between 1941 and 1944, she sank 37 Allied vessels; on 18 August 1944 depth charges dropped by a British Sunderland aircraft sank her in the Bay of Biscay.

U-107 (Type IXB)
Displacement: 1170 tonnes (1152 tons) submerged
Dimensions: length: 78.5m (258ft); beam: 6.7m (22ft)
Propulsion: surfaced: diesels 3281kW (4400hp); submerged: electric motors 745kW (1000hp)
Armament: 4 x 533mm (21in) torpedo tubes; 1 x 105mm (4.13in) deck gun; 2 x AA guns
Range: surfaced: 19,300km (11,992 miles) at 10 knots (18km/h; 11mph); submerged: 103km (64 miles) at 4 knots (7km/h; 5mph)
Speed: surfaced: 17.5 knots (32.4km/h; 20mph); submerged: 6.9 knots (13km/h; 8mph)
Crew: 48–56

out sustained long-term submerged operations with just occasional surfacing. They were the world's first genuine submarines rather than mere submersibles.

The Type XXI submarine displaced 1621 tonnes (1595 tons) when surfaced. The boat had two MAN diesel engines, two double-acting electric motors, two silent-running electric motors, and trebled battery capacity (as compared to the Type VIIC). This enabled the submarine to travel submerged at 5 knots (9km/h; 6mph) for up to 72 hours before having to recharge its batteries; the latter could be done via a Schnorchel tube in five hours.

The design could achieve a remarkable top underwater speed of 17.2 knots (32km/h; 20mph), or 6.1 knots (11km/h; 7mph) using just its silent-running motors. This submarine could achieve a maximum underwater range of 630km (390 miles) when moving at a constant 5 knots (9km/h; 6mph). Both the underwater speed and endurance of the Type XXI was astonishing in comparison to previous submarines.

In 1943–45, shipyards at Hamburg, Bremen and Danzig constructed 118 Type XXI boats. By 1 January 1945, the Navy had commissioned 62 Type XXIs, although these boats' crews subsequently needed to complete working-up training in the Baltic. This working up was severely hampered by Allied actions, including aerially delivered sea mines, repeated aerial attacks, and surface ship patrols. Thus, it was only during March 1945 that the first Type XXIs reached near-operational readiness.

With Anglo-Canadian ground forces advancing towards northern Germany, the Kriegsmarine decided that the U-boat bases in Norway, rather than those in northwestern Germany, were now the best locations from which to launch this new 'Electro-boat' offensive. During February–April 1945, therefore, the first six nearly-operational Type XXIs sailed from the Baltic to Norway's southern ports.

Even as the war entered its final few days, the Germans appeared to still be contemplating continuing military resistance from Norwegian soil to enable the 'Electro-boat' offensive to unfold. On

Schnorchel

Close-up of the *Schnorchel* breathing tube; during 1944-45 some 160 such tubes were fitted or retro-fitted to Class VIIC and VIIC/41 boats; the new generation of 'Electro-boats' all also sported the device.

3–5 May, for example, German warships, merchant vessels and submarines (including five almost-operational Type XXIs) departed Kiel and sailed to southern Norway. Luckily for the Allies, the war ended before any sustained offensive by numerous potent Type XXI boats could be unleashed.

TYPE XXI: FIRST COMBAT MISSION

On 30 April 1945, the day that Hitler committed suicide, U-2511 departed from Bergen to become the first Type XXI to undertake a combat mission. The boat undertook an uneventful four-day combat patrol before, on the afternoon of 4 May, she received the orders from the new German president, Grand-Admiral Karl Dönitz, that ended offensive U-boat operations against the Western Allies.

On its return voyage to Bergen, U-2511 claimed that it remained undetected while mounting a dummy attack on the British cruiser HMS *Norfolk*. The only other Type XXI to put to sea, U-3011, departed Wilhelmshaven on 3 May; she encountered no enemy vessels during this brief patrol before Dönitz halted the offensive U-boat war.

Type XXIII 'Electro-boat' U-boats

The other 'Electro-boat' design that the Germans developed was the small Type XXIII boat, optimized for short patrols with sustained submerged sailing in shallow coastal waters such as the North Sea and the Mediterranean. With a crew of 14–18 and displacing a mere 262 tonnes (258 tons) submerged, these streamlined-hulled submarines were so small that they could only carry one torpedo for each of their two tubes. This obviously limited their offensive potency; the torpedo even had to be loaded externally while in harbour.

With engines producing 432kW (580hp), these submarines could move underwater at an impressive 12.5 knots (23km/h; 14mph) and on the surface at 10 knots (18km/h; 11mph). However, these boats' surface operational range was limited to just 4200km (2610 miles) when cruising at a steady 8 knots

U-2511 (Type XXI)
Displacement: 1819 tonnes (1790 tons) submerged
Dimensions: length: 76.7m (252ft); beam: 6.6m (22ft)
Propulsion: surfaced: diesels 2985kW (4000hp); submerged: electric 3730kW (5000hp)
Armament: 6 x 533mm (21in) torpedo tubes; 4 x 20mm (0.79in) AA guns
Range: surfaced: 24,944km (15,500 miles) at 10 knots (18km/h; 11mph); submerged: 630km (390 miles) at 5 knots (9km/h; 6mph)
Speed: surfaced: 15.5 knots (29km/h; 18mph); submerged (maximum): 17.2 knots (32km/h; 20mph)
Crew: 57–60

U-2511
This cutaway of the Type XXI U-2511 shows at its front the six torpedoes carried for the three starboard-side bow tubes, as well as at the rear its diesel engines (aft of the conning tower) and its electric motors (further aft still).

(15km/h; 9mph); while underwater, the boat's maximum range was 310km (193 miles) when moving at 4 knots (7km/h; 5mph). The boat's maximum dive depth was 180m (591ft). Designed in 1942–43, 61 Type XXIIIs were commissioned during the war's final nine months.

Having completed most of their working-up training in the Baltic Sea, in late January 1945 the first six nearly operational Type XXIII submarines sailed from Kiel to the ports of southern Norway, where their training continued. In the ensuing weeks, another 11 near-operational Type XXIIIs arrived in the harbours of southern Norway, with a further three remaining in the German ports of Wilhelmshaven and Kiel. On 29 January 1945, U-2324 commenced the first Type XXIII combat patrol. In total during the war's final 13 weeks, six of these coastal 'Electro-boats' undertook nine combat patrols, spending a total of 204 patrolling days at sea. Three of these operational boats between them sank four Allied merchant vessels, amounting to 7510 tonnes (7392 tons).

None of these operational six boats were sunk by Allied action, but five others were. With the war ending before the Type XXIII offensive could commence on any scale, little was achieved despite the significant amounts of resources poured into the project. During early May 1945, the Germans scuttled 31 of the remaining 55 commissioned Type XXIII boats, many of them located near Flensburg.

Specialist U-boat types

The Germans constructed small numbers of other specialist submarine types. These included the eight large Type XB mine-laying submarines. Built

Type XXIII U-boat

In early February 1945 U-2326 joined active operations with the 11th Submarine Flotilla at Bergen in Norway. During 19–27 April 1945 the submarine mounted her one and only combat patrol, but failed to detect any targets before returning to Stavanger in Norway.

U-2326 (Type XXIII)

Displacement: 262 tonnes (258 tons) submerged

Dimensions: length: 34.68m (113ft 9in); beam: 3m (10ft)

Propulsion: surfaced: 1 × MWM RS134S 6-cylinder diesel engine, 575–630 metric horsepower (423–463kW; 567–621 shp)

Armament: 2 x 533mm (21in) torpedo tubes

Range: surfaced: 4200km (2610 miles) at 8 knots (15 km/h; 9 mph); submerged: 359km (223 miles) at 4 knots (7.4km/h; 4.6mph)

Speed: surfaced: 12.5 knots (23km/h; 14mph); submerged: 9.7 knots (18km/h; 11mph)

Crew: 14–18

U-459 (Type XIV)
Displacement: 1932 tonnes (1901 tons) submerged
Dimensions: length: 64.5m (212ft); beam: 5.9m (19ft)
Propulsion: surfaced: diesels 522kW (700hp); submerged: electric 306kW (410hp)
Armament: N/A
Range: surfaced: 19,874km (12,350 miles) at 10 knots (18km/h; 11mph); submerged: 56km (35 miles) at 4 knots (7km/h; 5mph)
Speed: surfaced: 14.9 knots (27.5km/h; 17mph); submerged: 6.2 knots (11.5km/h; 7mph)
Crew: 53–60

Type XIV
Commissioned during November 1941, the ocean-going supply submarine U-459 conducted six resupply patrols in the North and South Atlantic during 1942–43; she was sunk by Allied aircraft on 24 July 1943.

in 1939–44, these ocean-going boats carried 66 sea mines. Six of these eight boats were sunk by Allied action.

Another specialist design was the Type XIV 'Milk Cow' supply submarine, of which 10 were constructed. These submarines had no offensive weapons but carried fuel and supplies with which to resupply the combat boats. These boats carried out 36 missions, many of them off the US coast or in the Caribbean. The Germans also manufactured four Type VIIC U-Flak boats, equipped with numerous Flak guns, and intended to down low-flying Allied submarine-hunting aircraft. Finally, in 1943–44, the Germans built eight Type XVIIA and XVIIB 'Walter' Perhydrol-fuelled submarines that were used for trials. These boats, featuring advanced turbine propulsion, had an amazing maximum submerged speed of 24 knots (44km/h; 28mph), but due to massive fuel consumption these boats had very limited operating ranges.

Specialist U-boats

Specialist boat	Function	Submerged displacement
Type XB	Mine-laying	2753 tonnes (2710 tons)
VIIF	Torpedo transport	1200 tonnes (1181 tons)
XIV 'Milk Cow'	Resupply	1932 tonnes (1901 tons)
XVIIA, XVIIB	Experimental Walter-propulsion combat boats	A: 314 tonnes (309 tons) B: 342 tonnes (337 tons)

BIBLIOGRAPHY

Blair, Clay: *Hitler's U-Boat War* (2 vols.). London: Orion, 2000.

Davison, Donald (ed.): *Warplanes of the Luftwaffe*. London: Grange, 1991.

Diedrich, Hans-Peter: *German Jet Aircraft 1939–45*. Atglen, PA.: Schiffer, 1995.

Edwards, Roger: *Panzer: A Revolution in Warfare, 1939–1945*. London: Arms and Armour Press, 1989.

Engelmann, Joachim: *German Artillery in World War II*. Atglen, PA.: Schiffer, 1995.

Frank, Hans: *S-Boats in Action during the Second World War*. London: Pen & Sword, 2007.

Grove, Eric: *World War II Tanks: The Axis Powers*. London: Orbis Publishing, 1971.

Hogg, Ian: *German Artillery of World War II*. London: Arms & Armour Press, 1975.

Mallmann Showell, Jak P.: *Hitler's Navy: A Reference Guide to the Kriegsmarine 1935–45*. Annapolis, Md.: Naval Institute Press, 2009.

Milsom, John: *German Military Transport of World War II*. London: Arms & Armour, 1975.

Phillpott, Bryan: *The Encyclopedia of German Military Aircraft*. London: Bison, 1981.

Walter, John: *Guns of the Third Reich: Small-arms of Hitler's Armed Forces, 1933–45*. Stroud: The History Press, 2016.

Whitley, M.J.: *German Capital Ships of the Second World War*. London: Cassell, 2001.

Williamson, Gordon: *German Destroyers 1939–45* (New Vanguard Series). Oxford: Osprey, 2003.

Williamson, Gordon: *German Light Cruisers 1939–45* (New Vanguard Series). Oxford: Osprey, 2003.

INDEX

References to photographs are in *italics*.